VD - 14

Biogenesis
Evolution
Homeostasis

A Symposium by Correspondence

Editor
A. Locker

With 13 Figures

Springer-Verlag
Berlin Heidelberg New York 1973

ISBN 3-540-06134-7 Springer-Verlag Berlin Heidelberg New York
ISBN 0-387-06134-7 Springer-Verlag New York Heidelberg Berlin

This work is subject to copyright. All rights are reserved, whether the whole or part of the material is concerned, specifically those of translation, reprinting, re-use of illustrations, broadcasting, reproduction by photocopying machine or similar means, and storage in data banks.

Under § 54 of the German Copyright Law where copies are made for other than private use, a fee is payable to the publisher, the amount of the fee to be determined by agreement with the publisher.

The use of registered names, trademarks, etc. in this publication does not imply, even in the absence of a specific statement, that such names are exempt from the relevant protective laws and regulations and therefore free for general use.

© by Springer-Verlag Berlin · Heidelberg 1973. Library of Congress Catalog Card Number 72-96743. Printed in Germany.
Offsetprinting and bookbinding: Julius Beltz, Hemsbach/Bergstr.

Preface

In 1963, 1965 and 1967 "Symposia on Quantitative Biology of Metabolism" were organized on the pretty island of Helgoland (Biologische Anstalt) by me in collaboration with O. KINNE and F. KRÜGER. Unfortunately this worthy approach towards bringing together interested scholars in a regular way ceased, mainly for financial reasons, although the need for and interest in conferences like these for the exchange of ideas on special topics unchangeably persists. So I had to look for other possibilities and one of them was to try to arrange a similar conference under the auspices of NASA. This institution, however, eventually retreated, although during my discussion with its representatives a new special theme gained shape. This is the topic to which this volume is devoted: Biogenesis. It is also treated in a new way that probably could be a model for similar undertakings: a symposium by correspondence. In this new approach to scientific information exchange, manuscripts were collected by the editor and sent to every contributor for his comments. The author could then reply by means of a concluding remark. So, in many instances very valuable ideas concerning the topics of the several papers could be gathered and it is to be hoped that this procedure has conferred upon this volume a certain uniqueness. Of course, I had to observe the agreement with Springer-Verlag not to exceed the extent of the volume allotted to us. I was, therefore, sometimes placed in the predicament of having to shorten the submitted paper and in other cases I was forced to accept only the abstract of the paper. In these instances I must ask the authors for their understanding and patience, since many of these papers will presumably be published in a forthcoming journal.

I take the opportunity to thank all the contributors to this volume for their valuable cooperation which made it possible to complete a work which deals indeed with a theme of outstanding topicality. The Austrian Society for Atomic Energy, Ltd, assumed the cost for distributing the many copies of the papers by air mail, thus enabling the exchange of the comments. I have especially to thank Dr. K. F. SPRINGER of Springer-Verlag for agreeing to publish this volume. My particular gratitude is directed to my friend, N. A. COULTER JR., who kindly enabled the completion of the editorial work during my stay with him under an NSF Senior Fellowship. My former secretary, Mrs. H. SONNECK, assisted me at the beginning of the editorial work.

It is hoped that this volume will arouse interest among that scientific community which is convinced that Theoretical Biology has now achieved the stage of scientific development characterized not only by the application of formal models but also by the use of epistemology in penetrating into the depth of the problems.

Vienna and Chapel Hill, March 1973 　　　　　　　　　　　　　　　　　　　A. LOCKER

Contents

I. General Formal and Relational Aspects

    A. Locker: Systemogenesis as a Paradigm for Biogenesis 1
    H. P. Yockey: Information Theory with Applications to Biogenesis and Evolution 9
    P. Decker, A. Speidel, and W. Nicolai: On the Origin of Information in Biological Systems and in Bioids (A) 25
    J. R. Hamann and L. M. Bianchi: On the Evolutionary Origin of Life and the Definition and Nature of Organism: Relational Redundancies (A) 27

II. Optimization and Evolution

    H. J. Bremermann: On the Dynamics and Trajectories of Evolution Processes 29
    D. Cohen: The Limits on Optimization in Evolution 39

III. Control and Homeostasis

    H. H. Pattee: Physical Problems of the Origin of Natural Controls . . 41
    Ch. Walter: The Significance of Cooperative Interactions in Biochemical Control Systems (A) 51
    B. Hess: Organization of Glycolysis (A) 53
    Z. Simon: Cell Models and the Homeostasis Problem (A) 55
    N. A. Coulter, Jr.: Contribution to a Mathematical Theory of Synergic Systems 57

IV. Oscillation, Excitability and Evolution

    B. Gross and Y. G. Kim: The Role of Precursors in Stimulating Oscillations in Autocatalytic Diffusion Coupled Systems 63
    S. Comorosan: Oscillatory Behavior of Enzymic Activities: A New Type of Metabolic Control System (A) 71
    Th. Pavlidis: The Existence of Synchronous States in Populations of Oscillators 73
    R. Wever: Reactions of Model-Oscillations to External Stimuli Depending on the Type of Oscillation (A) 81
    N. W. Gabel: Abiogenic Aspects of Biological Excitability. A General Theory for Evolution 85

V. Statistics and Thermodynamics

P. Fong: Thermodynamic and Statistical Theory of Life: An Outline . 93
H. C. Mel and D. A. Ewald: Thermodynamic Potentials and Evolution towards the Stationary State in Open Systems of Far-from-Equilibrium Chemical Reactions: The Affinity Squared Minimum Function (A) 107
G. Nicolis: Thermodynamic Stability and Spatio-Temporal Structures in Chemical Systems (A) 109
D. Detchev and S. Teodorova: Optimal Adaptation of the Metabolic Processes in the Cell (A). 111

VI. Metabolic Evolution

R. Rosen: On the Generation of Metabolic Novelties in Evolution . . 113
D. C. Reanney: Circular Nucleic Acids in Evolution (A) 125

VII. Time and Evolution

B. Günther: Physiological Time and Its Evolution 127
E. W. Bastin: Timeless Order 137

VIII. Learning, Memory and Evolution

A. M. Andrew: The Ontogenesis of Purposive Activity 147
J. S. Griffith: Some General Problems of Memory 159
E. G. Brunngraber: Role of Glycoproteins in Neural Ontogenesis, Membrane Phenomena, and Memory (A) 165
D. L. Szekely: On Controlled and Totally Neural-Replies Generated Concepts for Biology and Functional Brain Theory 169
R. E. Kalman: Remarks on Mathematical Brain Models 173

IX. Conclusion

A. Locker: How to Conceive of Biogenesis (A Reflection Instead of a Summary). 181

Subject Index 185

(A): Abstract

List of Participants

ANDREW, A. M., Department of Applied Physical Sciences, The University of Reading, Building 3, Early Gate, Whiteknights, Reading/Great Britain
BASTIN, E. W., Language Research Unit, 20 Millington Road, Cambridge/Great Britain
BREMERMANN, H. J., Department of Mathematics, University of California, Berkeley, CA 94720/USA
BRUNNGRABER, E. G., Research Department, Illinois State Psychiatric Institute, 1601 West Taylor Street, Chicago, IL 60612/USA
COHEN, D., Department of Botany, The Hebrew University, Jerusalem/Israel
COMOROSAN, S., Department of Biochemistry, University Postgraduate Medical School, Fundeni Hospital, Bucharest/Romania
COULTER, N. A., JR., Curriculum in Biomedical Engineering and Mathematics, UNC Medical School, University of North Carolina, Chapel Hill, NC 27514/USA
DECKER, P., Chemisches Institut, Tierärztliche Hochschule, 3 Hannover, Bischofsholer Damm 15/W. Germany
DETCHEV, G., Central Biophysical Laboratory, Bulgarian Academy of Sciences, Sofia/Bulgaria
FONG, P., Physics Department, Emory University, Atlanta, GA 30322/USA
GABEL, N. W., Illinois State Psychiatric Institute, 1601 West Taylor Street, Chicago, IL 60612/USA
GRIFFITH, J. S. (deceased; formerly): Department of Chemistry, Indiana University, Bloomington, IN 47401/USA
GROSS, B., Research Department, Mobil Research and Development Corp., Paulsboro, NJ 08066/USA
GÜNTHER, B., Departamento de Medicina Experimental, Universidade de Chile, Casilla 16038, Santiago/Chile
HAMANN, J. R., Center for Theoretical Biology, SUNY at Buffalo, 4248 Ridge Lea Road, Amherst, NY 14226/USA
HESS, B., Max-Planck-Institut für Ernährungsphysiologie, 46 Dortmund, Rheinlanddamm 201/W. Germany
KALMAN, R. E., Department of Mathematics, Stanford University, Stanford, CA 94305/USA
LOCKER, A., Österreichische Studiengesellschaft für Atomenergie GmbH., Lenaugasse 10, 1082 Wien/Österreich (jointly 1972-1974; Curriculum in Biomedical Engineering and Mathematics, UNC Medical School, University of North Carolina, Chapel Hill, NC 27514/USA)

MEL, H. C., Donner Laboratory of Medical Physics, University of California, Berkeley, CA 94720/USA
NICOLIS, G., Faculté des Sciences, Université Libre, Bruxelles/Belgique
PATTEE, H. H., Center for Theoretical Biology, SUNY at Buffalo, 4248 Ridge Lea Road, Amherst, NY 14226/USA
PAVLIDIS, T., Department of Electrical Engineering, Princeton University, Princeton NJ 08540/USA
REANNEY, D. C., Lincoln College, Canterbury/New Zealand
ROSEN, R., Center for the Study of Democratic Institutions, Santa Barbara, CA 93102/USA
SIMON, Z., Catedra Chimie Organica, Universitatea Timisoara, Bd. Parvean 4, Timisoara/Romania
SZEKELY, D. L., Association of Unification and Automation in Science, P.O.B. 1364, Jerusalem/Israel
WALTER, CH., Department of Biomathematics, M. D. Anderson Hospital, University of Texas, Houston, TX 77025/USA
WEVER, R., Max-Planck-Institut für Verhaltensphysiologie, 8131 Erling-Andechs/W. Germany
YOCKEY, H., Atomic Pulse Radiation Facility, Aberdeen Proving Ground, MD 21002/USA

Systemogenesis as a Paradigm for Biogenesis

A. Locker

Abstract

Starting with a description of the features of a system as a given entity it is shown that an adapting or self-organizing system undergoes transformations that can be extended to formally explain self-generation. This process depends on the interplay between structurally determined constraints and relationally enabled decisions which a "self"-like representation of the system can use to realize itself and thus to constructively build up transitions of states with ever increasing complexities. The need of a certain a priori framework for those considerations is outlined.

I. The Problem

It seems to be easier to grasp and understand the being as such than to find a reasonable explanation for its origin. The difficulties in assigning an origin to ordered phenomena, called systems, are widely accepted. In tackling the problem it may possibly be wise to look at these properties of systems which could offer a hint at the commencement of their existence. A further step in approaching the problem could be furnished by cogitating about the means systems have in re-organizing themselves following perturbing influences or in adapting against new situations. However, for the proper understadding of the genesis of a system, i.e. an organized complex emerging out of unorganized entities, a certain reorientation of thinking is required. In this reorientation the problem of measurement (or of the observer) is of paramount importance. It intimately relates to the problem of how to appropriately apply a conceptual framework (or a mode of consideration) to a description.

Based upon the recognition of these epistemological aspects an attempt is undertaken to comprehend biogenesis as a specification of the more formally accentuated problem of systemogenesis. The reference to the conscious subject possibly enables one to outline systemogenesis in a constructive (or pragmatic) way.

II. System as Given Entity

All the definitions of systems proposed up to now have in common that they implicitly contain certain definitions of structure, order, complexity or the like. These implicit definitions presuppose, therefore, the existence of interrelated parts or system variables. Parts (elements) can be defined qualitatively, variables quantitatively. Since, however, a system is characterized also by its activity or function, it is possible to ascribe activity to the parts (elements) as well as the relations connecting them. The elements exert their effects onto their proper environment, e.g. other parts (elements) in the system. Activity is thus channeled and guided by the relations. It is the relations which allow to distinguish the system from a pure collection of individual elements isolated from each other. In dependence upon the kind of relations, e.g. whether they are permanent or temporary, strong or weak, etc., certain alternatives of activity are given. By means of these alternatives certain possibilities of activity can be delineated. Consider (2) a product space, i.e. the space of possibilities, within which some sub-set of points indicates the actualities. In order to make the distinction of (or transition from) possibilities and/to actualities, some constraints, e.g. boundary conditions, must become manifest. Therefore, one is entitled to discern an S-Structure, dealing with the invariant, time-independent relations (couplings) between the elements from the P-structure, denoting the timely behavior or activity (8). By the term P-structure also the transition is expressed that occurs according to a certain program between several states of the system.

Intimately involved in handling the interplay between analysis and synthesis is the identification problem. New qualitative methods have recently been proposed which circumvent quantitative limitations and are applicable also to instances of unknown initial conditions (3). The whole task of parameter estimation must be subsumed under the problem of the recognition of the system's organization. Besides this, it must be decided what has to be done with the identified parameters.

The approach to organization or order is opened up by additional specifications. It has been suspected quite correctly, that order is so-to-speak projected by the observer into the system (10) and its extent evidently depends on the subtlety of the mode of description applied. Two extreme cases can in principle be conceived of, namely one according to maximal simplicity and the other according to maximal complexity. Both must interact in allowing the adequate system description. A simple system would be entirely irredundant, i.e. no aspect of the system can be derived from any other aspect (19). However, also a totally complex system is epistemologically trivial (10), in that it exhibits constraintless interaction of the elements in all directions, which is nothing but another kind of simplicity. The appropriate description is furnished only by applying the relevant constraints. The interaction between simplicity and complexity in the system's description is determined by optimality considerations, e.g. with respect to stability problems. Optimality considerations result, e.g., in the statement that a loose coupling is optimal (2) (13). Zero coupling means zero organization, whereas maximal organization requires a strong coupling, which possibily can lead to the loss of identity of the elements, the system thus being replaced by a field (10).

However, as far as one is induced to assume the existence of sub-systems (regions) within a system, also called aggregates, clusters or multi-element components, it becomes necessary to distinguish between the 2 kinds of interrelation, that now appear: 1. Within the components the coupling of the elements is commonly assumed to be strong, 2. whereas between the components the interactions are weak (19). This distinction can force the observer to make an additional statement, namely to ascribe to the system a hierarchical order. The transition between alternative states of a system exhibiting the system's activity, can generally be regarded as a mapping of one-to-one (8), occurring only on one level of the system's description, so-to-speak horizontally. In contrast to this, a series of vertical many-to-few mappings creates a hierarchy. During the transcending process to the next level the number of parameters needed for one level is reduced. This constructive process entitles the observer to leave the view of the system as a given one behind and to enter into a consideration that deals with the system's capability to actively increase its diversification.

III. System as Adapting and Self-organizing Entity

In as much as the system is intimately linked with its environment, the inner dynamic forces of the system constantly change. If the global activity of the system is to be maintained the change of the system must occur in such a way as to eventually enable again this activity. Such a kind of change is called adaptation. Several definitions of adaptation have been given (18), either 1. with regard to the maintenance of the original system's activity or 2. with regard to an improvement of this activity, i.e. an increase of its efficieny. Into each definition the elements and the relations of the system, determining the system's activity, are involved.

The influence which elicits an adaptation can generally be considered as a stimulus (i.e. perturbation) and the adaptation as a response to it (i.e. stabilizing process). Dependent upon whether the stimulus is located within or outside the system, one can delimit internal adaptation from an external one. In order to adapt the system must either modify itself or alter the environment or both. It is possible to classify the system's behavior during the adaptation process into 1. goalseeking, 2. purposive and 3. purposeful systems. Goalseeking adaptation consists of a one-to-one correspondence (i.e. mapping) between the stimulus and the response, that is to say, to each stimulus only one response belongs. In contrast to this, the purposive system displays a one-to-many correspondence, which means, that each stimulus can elicit more than only one response. By the term purposeful system the system's behavior is being based, during the process of adaptation, on a change of the system's function. Thus, if the system's original function consisted in producing a certain activity, then, after the accomplishment of adaptation, the system exerts another kind of activity. Any system organization which changes its objectives in responding to any type of stimulus pertains to this particular category of adaptation, which is typically a functional adaptation.

Since the system achieves, during the process of adaptation, a new kind of order (or organization), be it structural or functional, it undergoes self-organization. This process includes not only adaptation as such, but also cognition (of new inner or outer situations, against which the system has to adapt) and learning. Especially cognition implies the interaction with surroundings that must be recognized by means of information obtained from them; therefore, the subjective aspects cannot be neglected here, i.e. those of observer and measurement. The system behaves like an entity with a subjective self, in that it filters out of the bulk of information it generally receives that essential part which it can use for a change of its behavior. This automatically leads to the statement that an observer, behaving similarly to ourselves, controls the system. Within the control mechanism (i.e. within the observer) there must be a representation of the controlled entity. For characterizing the self-organizing system several attributes can be chosen. Whatever these attributes be, their number is assumed to increase as a result of the interaction between the controller and the controlled system. This assumption obeys PRINGLE's criterion (16), stating that during the process of self-organization more independent variables are required to describe the system at a later instant than at an earlier one. This is simply a consequence of the fact that not only more elements are included but also that the manner of interaction between these elements changes, i.e. the complexity increases.

In addition to the constraints mentioned above a self-organizing system requires further definite constraints. Especially with these constraints it becomes obvious that they are structural, i.e. involving matter and energy. In a model proposed (14) it has been assumed that the constraints determine the transfer of energy that need take place from the environment to the system. Depending on the ratio of the inflow rate of energy to the outfow rate a nucleation (i.e. clustering or aggregation of elements thus forming a region) may occur. In these regions a conversion of energy into signal is carried through. This is a principal statement connecting thermodynamics with information theory (9) (21). The activity of the energy converters is decisive for the signal connectivity which in turn defines the relations of interaction between the elements. With further increase of the energy inflow - depending on the relevant constraints - an irreversible change can take place corresponding to the idea that too densely or too loosely packed a system will inextricably tend to instability and finally to "attrition". This view is again a subjective one. It is only the observer who partitions the system into regions, whereby a many-to-few mapping is pursued.

In that the regions are characterized by their states, the process of self-organization can be understood as a transition between these states undergoing new formations of the connections that exist between the elements provided the stimuli (i.e. perturbations) exceed certain threshold values. The possibility for those transitions implies a certain lability of the connections. It implies also the build-up of hierarchies.

IV. System as Self-Generating Entity

The main problem is how the system is able to generate itself, therefrom exceeding a pure re-adjusting and self-organizing process. The model just outlined regards as fundamental for the process of adaptation and self-organisation a clustering and forming of regions within a system. In the generation of the system itself one can easily conceive of a certain similarity with this mechanism. If through an inflow of energy into an equally distributed pre-system stages (or transitions) like those supposed during the process of self-organization will be passed through, such that clusters (or regions, etc.) are formed, then we have already a certain generation of system before us. Such a transition is qualitative, so that prior to it we are practically not dealing with a system, but rather with a kind of pre-system, and after it we are confronted with an entity that actually deserves the term system.

A presupposition for this process is that the qualitative transition brings about an interaction between the elements in contradistinction to lacking interaction in the unorganized pre-system (19). At the same time this interaction has to assume a certain form. Namely, it has been repeatedly demonstrated (7) (11) (15) that in the particular transition of states such that a hierarchical order arises PRINGLE's criterion does not hold. Instead, with increasing complexity the description becomes simpler. This is a striking consequence of the observer's role in the description be it an outside observer or an observer-like representation within the system. The observer uses fixed structural constraints as a record of the behavior etc. of the elements below a certain level, whose complexity reappears as unified, i.e. simplified above that very level. Therefore, a mapping is done somewhat similar to the human observer's semantic steps employed in the process of cognition. This many-to-few mapping depends on decision processes (4) stating a formal dynamics for the continuous construction of further levels. This consideration underlines again the inevitable need to introduce here terms that stem from subjective experience and without them it appears inappropriate to handle the problem of creating hierarchies and mastering the emergence of order.

By means of the observer-like record in the system, replacing so-to-speak a "self", the formal representation becomes a formal realization (4). This can be understood in the sense (as indicated by Fig.1) that both, representation as well as realization, are interrelated in a circular way. Hereby, reception, i.e. passive mapping, is tied to effectiveness, i.e. active mapping (and vice versa). In viewing these relations horizontally, as existing on a single level, a one-to-one mapping of the elements into themselves is self-representation and self-realization simultaneously. Regarded in a vertical respect the boundary case of the (many-to-few) mapping is two-to-one, allowing the representation of the self-represented lower level at the next higher level. However, this passive record is immediately converted into active control, if a one-to-two (or few-to-many) mapping ensues from a decision made on the higher level. By means of this decision the higher level realizes itself. Analogous processes occur as long as the diversity of the environment is recorded by the several kinds of mapping or, alternatively, the system responds to the environment in an active (i.e. adaptive) manner (as indicated above). These actions and re-actions, respectively, concern all the different organizational levels of the system. Thus, the observer-like "self" (or center) of the system is created by the help of an interplay between structurally determined constraints and relationally enabled decisions jumping to and fro between representation and realization. The constructive aspect seems to be guaranteed by the endless continuibility of this self-creation, being at the same time the system's creation.

The relational framework for decisions is an a priori one in that it provides the presupposition for constraints which in turn exhibit the structural concretizations and, hence, those compulsory restrictions whose "empty space" only entails function, i.e. the activity of the system. It constitutes also the independent variables. The a posteriori aspect assigns numerical quantities to the elements perceived or defines the dependent variables (12). Build-up of higher forms of order (like hierarchies) implies both aspects in that complexity increases, but the purely metric aspect has to be embedded into the relational one, stating that despite this increase some other features, namely that of order, have to remain invariant. If they would not remain constant, thus constituting a framework, it would become impossible to assess the increase in complexity.

The framework of order must not be necessarily considered as purely static. In discerning a dynamic view of order, implying dissolution of an older decrepit order and origination of a newer one, that means ordering from order, it is immediately evident that order has intrinsically to do with processes (5).

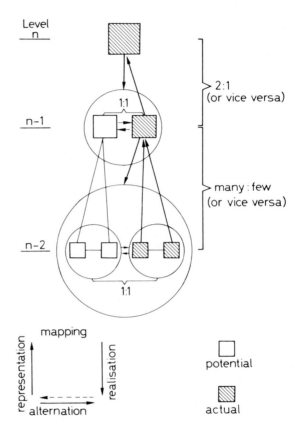

Fig.1. Connection of the circular way: representation-alternation-realisation with the system's hierarchy

The discarding of the limits of automorphism, i.e. a restricted field that allows the assumption of different states within it, opens up true generability of the new, i.e. true creativity (5) or change of order. But there must be supposed something as fundamental beyond this change: either a kind of tension between harmony and conflict or a kind of transformation from the fortuitous to the necessary (or vice versa). The last mentioned distinction corresponds to F- and L-truth in logic. It must not be overlooked, that although both sides belong together, the supplementary third can again be replaced by the observer, who, in the process of uniting these contrasts (or complementarities), assumes the active position of a decision-making "self". In performing creation (or generation) it uses two different time scales: 1. one which enables the consideration of the evolution of structure in a horizontal way, i.e. within one level and 2. one which enables the vertical transgression between the levels. The former time scale exceeds by far the latter, which in its extreme case has no metric, i.e. the (at least cognitive) transgression is instantaneous. Because hierarchical order enables the building (i.e. ordering) of hierarchical structures, the former is in itself a time-less presupposition for the latter. Those time scales can be elicited, which can be used, so-to-speak as operative times, for performing this ordering.

V. The "Myth" of the Origin of Life

As is well known the probability for spontaneous formation of a living organism out of a pre-biological soup is extremely low; the probability increases only if, in addition, the property of self-multiplication is assumed. Granting this, it is not difficult to conceive of kinetic mechanisms, that could play a role in the evolution on the molecular level (17). Apart from autocatalysis and reflexive catalysis, the formation of cyclic reaction systems and of regulation systems have been regarded as fundamental. Starting from these considerations, quite a story of the origin of life as well as of the evolution of organization can be told.

However, students of microevolution in bacteria etc. stress the fact that there is no continuous gain of new features during evolution, but instead evolution is characterized by a gradual loss of potentialities - an experimental view which recently has been confirmed from the standpoint of the theory of structural stability (20). In generalizing this fact it seems to indicate that any gain has to be paid for by a loss or, as in ancient myths is narrated, creation is enabled by a victim (6). The inevitable involvement of the human subject with its deep internal experience, seems to point out that only the transformation of a "profane" space into a "sacred" (transcendental) and of a concrete time into a mythical (narrative) provide the conceptual framework for a proper understanding of the problem. In the foregoing discussion, which followed BOHM's "metaphysics of process" (5) in that the three chapters outlined correspond 1. to quasiequilibrium processes, 2. to dynamical processes and 3. to creative processes, nothing more has been done than trying to conceptually repeat creation of life in order to perpetually renew it by making conscious of the unperishable congruity of the subject's life with the so-called objective life.

This touches upon the question whether life can only be quasicreated by transmission from a lively organism to pre-biotic entities or whether it generates itself by evolving (or emerging) out of the ensemble of pre-biotic entities. This question must remain freely pending so long as one is not willing to definitely decide on one or the other of the two irreconcilable standpoints.

References

1. ACKOFF, R.L.: Choice, Communication and Conflict. Philadelphia: Management Science Center. 1964.
2. ASHBY, W.R.: Design for a Brain. New York: John Wiley 1952.
3. BELLMAN, R.: Math. Biosci. 5, 201 (1965).
4. BIANCHI, L.M., HAMANN, J.R.: J. Theor. Biol. 28, 289 (1970).
5. BOHM, D.: In: WADDINGTON, C.H. (Ed.): Towards a Theoretical Biology, 2. Sketches, 18, 41, Edinburgh Univ. Pr. 1969.
6. ELIADE, M.: Kosmos und Geschichte, Hamburg: Rowohlt (rde) 1966.
7. HIGGINS, J.J.: In: CHANCE B., ESTABROOK R.W. & WILLIAMSON J.R. (Eds.): Control of Energy Metabolism, p.13, New York-London: Academ. Pr. 1965.
8. KLIR, G.J.: Gen. Syst. 13, 13 (1968).
9. LAMB, G.G.: Gen. Syst. 13, 165 (1968).
10. LEVINS, R.: In: WADDINGTON C.H. (Ed.): Towards a Theoretical Biology, 3, p. 73, Drafts, Edinburgh Univ. Pr. 1970.

11. LOCKER, A.: Rev. Biomath. (Paris) 8, 87 (1970).
12. McKAY, D.M.: quoted from (14).
13. MOROWITZ, H.J., HIGINBOTHAM W.A., MATHYSSE S.W., QUASTLER, H.:
    J. Theor. Biol. 7, 98 (1964).
14. PASK, G.: Gen. Syst. 4, 151 (1959).
15. PATTEE, H.: In WHYTE L.L., WILSON A.G. & WILSON D. (Eds.):
    Hierarchical Structures, Elsevier, New York 1969, p.161.
16. PRINGLE J.W.S.: Gen. Syst. 1, 9o (1956).
17. QUISPEL A.: Acta Biotheor. 18, 291 (1968).
18. SAGASTI, F.: Gen. Syst. 15, 151 (1970).
19. SIMON, H.A.: Gen. Syst. 1o, 63 (1965).
20. THOM, R.: In: WADDINGTON, C.H. (Ed.): Towards a Theoretical Biology,
    1. Prolegomena, Edinburgh Univ. Pr. 1968, p.152.
21. TRIBUS, M.: Gen. Syst. 6, 127 (1961).

DISCUSSION

BREMERMANN:
I greatly enjoyed reading this paper and I find parts 1 through 4 contain many ideas that coincide with my own understanding of systems. I am afraid however, that this paper may draw some criticism for being too abstract. Some applications or concrete examples would make the paper much more accessible and more meaningful to the average biologist. Part 5 I found hard to understand.

LOCKER:
Although some instructive examples of possible realizations of self-organizing and learning systems have been given (e.g. BARRON, 1966, and some other papers in YOVITS et al., 1962, OESTREICHER-MOORE, 1966, and v. FOERSTER et al., 1969) we have to admit that generally all these approaches are so abstract that the gap between theoretical considerations and practical applications has not yet been bridged. But the question must be raised whether this gap, although seemingly contradicting the McCULLOCH-PITTS theorem, is inevitable. I surmise that it is an expression for the neglect of necessarily dealing with relations and with the context that determines their significance. Up to now systems seem to have been implemented as to act only according to context free data (DREYFUS 1972). For further problems, also that touched upon in item 5, I may refer to my concluding remark to this symposium and to a forthcoming paper (LOCKER & COULTER Jr., 1972 Int. Cybernetics Conference, Washington, October 1972).

MEL:
The GARDNER and ASHBY demonstration (Nature 228, 784 (1970)), that too high a degree of connectivity leads to system instability, may be a quantitative statement of (part of) the idea - outlined in part 3 - i.e. be more objective than subjective.

LOCKER:
I am glad of this remark. Indeed, what at first sight appears as objectively occurring can be understood only by making reference to our subjectivity, i.e. the consciousness, by making it analogous to ourselves.

SZEKELY:
What characterizes a living species is a very high-level concept of functional invariance: a sequence of successively subordinated concepts of invariance, functional-dynamic at each level and in their totality. LOCKER states that the relation between the higher and several lower-level invariant subsystems is in general of the form of a many-one relation. I would like to complement this by SOMMERHOFFS "directive correlation" and with reference to the analysis of H. REICHENBACH in "Die Axiomatik der Einstein'schen Raum-Zeit Lehre" to use instead of "directive correlation" the form "directed coordinative relation". – Each species is a variant of a high type invariant. Phylogenesis is the genesis of invariances that transgress under transformation their characteristic unchanged logical structure. The more the concepts of taxonomy encompass invariances of only a few structural characteristics, the more include the specific subdivisions detailed structural invariance concepts. But all these forms of invariances are anchored genetically by a very detailed state of the genetic code. Phylogenesis refers from a macroscopic viewpoint to invariances, but from the point of view of mechanisms and control actions at the molecular level to a sequence of states of the code. – I fully agree with LOCKER's emphasis on relational bases for thinking about such problems and with the categorization of stimuli in the process of adaptation. He deals with adaptation having phylogenetic influence, i.e. from the transition from one invariance to another under the pressure of the environment. Self-organization must be expressed in state-alterations of the control-code. – Regarding the assertion in §.III "It is only the observer who partitiones the system into regions" I can accept this for a totally abstract and information-theory oriented approach. But biosystems are physical and the subordinate sequence of invariant forms and functions makes subdivisions of formal nature redundant. The stimulus affects in some way the totality of the system. The higher it reaches along the organizing hierarchy, the smaller its direct influence, the longer the process of collection and integration to reach the threshold: that threshold which is enforcing a structural change in the state of the control-code and along with this in the structure of the invariance.

LOCKER:
If you would not divide you would not recognize (compare GOETHE's "Trennen und Verbinden"). But these operations occur only in the formal (and cognitive) sphere, otherwise you would not be able to recover the separated entities and thus understand, in analogy to yourself, how things work. – It is correct to say that the stimulus response mechanisms depend on the system's organization. Actually, it is established in such a way that at the lowest, i.e. the most peripheral, level redundancies occur, whereas most difficult activities, like self-repair, are placed on the highest (or most central) level (CHENG, 1964).

Information Theory with Applications to Biogenesis and Evolution

H. P. Yockey

Abstract
The representation theorem and the channel capacity theorem of information theory are applied to the molecular biology of biogenesis and evolution. From the information theoretic point of view it is essentially incredible for a particular given protein to arise de novo by chance alone. Genetic information could arise in a series of steps in which a useful sequence of amino acids or a sequence of improved specificity arises by chance. Such an improvement will be incorporated in the genome in a time short compared to the time required to appear by chance. One-to-one codes are part of the modern protein synthesis and may have been used by the most early form of life. The environment would have been modified by the generation of new substances. A one-to-one code may have led to a binary code and then to the modern triplett code. At each step the primitive life may have modified its environment irreversibly. Proteins with specificity may form a chain so that one is transformed to another by a series of amino acid substitutions, additions or deletions. There may be many such paths of specificity.

I. INTRODUCTION

The concepts of information theory have been important in biology since at least 1953 when WATSON & CRICK brought to a focus the informational role of DNA in molecular biology. That something more abstract than a simple template is involved is shown by the fact that DNA is composed of four elements and protein is composed of twenty. GAMOW (2) was quick to point out that something rather like a Morse Code and a letter alphabet was involved.

We cannot expect to present a full treatment in the pages allotted. The reader must be content to be referred to other works for amplification and for the more detailed and difficult proofs; we can, however, discuss the meaning of the theorems and correct some wrong impressions. In our explanation of information theory and its use in biology, we will adhere as closely as possible to the storage of genetic information in DNA and to the transcription information processes of protein formation. This will lead us to some interesting problems in biogenesis and evolution.

The notion that evolution proceeds by random mutation and natural selection is one of the great principles of modern biology. Life, therefore, seems to flout the second law of thermodynamics. Biological organisms seem to be something more than chemical systems yet at the same time measurements on these systems reveal that natural laws are obeyed. The fact that information is at the same time a quantity which can be defined mathematically and operationally and yet exists in living matter but not non-living matter may perhaps contribute to a resolution of this paradox.

The reason that genetic messages can be recorded in informational biomolecules is that the order of the component elements (i.e., nucleotides or amino acids) is not determined by a minimum in the potential energy, the free energy or by some purely chemical factor. A crystal of sodium chloride is orderly but it can store no information because this orderliness is determined by the potential energy pattern of the crystal. Non-living matter never stores information by itself, although information may be stored by outside action (i.e., photographic plate, etc.). Perhaps the ability to store, transcribe, duplicate and generate information is the difference between living and non-living matter.

II. A SKETCH OF THE IDEAS OF INFORMATION THEORY AS APPLIED TO BIOLOGY

First, we must establish a definition of information and its quantitative measure. Consider a section of DNA constituting a cistron. The nucleotide bases are ordered in a special way. The special sequence is a selection from all possible arrangements and we shall call any such sequence a genetic message if it contains biological specificity. There is redundance +) in the code and furthermore some choice or uncertainty in the amino acid selection (12) is allowed without appreciable loss of specificity. Because of this variation we will have an ensemble of genetic messages which will represent the genetics of the ensemble of organisms including all possible mutations (25). We will take this point up later.

Let x be a variable which characterizes the selection of nucleotides or amino acides in an informational molecule. Let x_i characterize a particular choice. We can then speak interchangeably for any informational molecule. Associated with each x_i there will be a probability p_i and given all p_i there will be a probability distribution.

We wish to establish a measure of choice or uncertainty, $H(p_i)$, in the selection of the x_i. Probability distributions which are sharply peaked reflect little choice or uncertainty; those which are broad show a maximum of choice or uncertainty. One of the great advances which was made by SHANNON (20) lies in the discovery that there is a unique, unambinguous, criterion for the amount of choice or uncertainty reflected by a probability distribution. In our discussion "choice" pertains to the sender or source of information and "uncertainty" to the receiver or destination. The following conditions seem to be reasonable requirements for such a measure, $H(p_i)$.

1. H is a continuous function of the p_i
2. If all $p_i = 1/n$ then $H(1/n, 1/n...1/n)$ is a monotonically increasing function of n.
3. If a choice is composed of two successive choices, the original H is the weighted sum of the individual values, H_i. If

$$\omega_1 = \sum_{i=1}^{k} p_i \; ; \; \omega_2 = \sum_{k+1}^{m} p_i, \text{ etc.}$$

$$H = H(\omega_1, \omega_2) + \omega_1 H \left(\frac{p_1}{\omega_1}, \frac{p_2}{\omega_1}, \ldots \frac{p_k}{\omega_1}\right)$$

$$+ \omega_2 H \left(\frac{p_{k+1}}{\omega_2} + \ldots \frac{p_m}{\omega_2}\right) + \ldots$$

The three conditions appear at first sight rather bland and not very demanding. They are enough, however, to determine uniquely the form of the measure, H, namely.

+) This is called degeneracy in molecular biology but redundance is a better term.

(1) $$H(p_i) = -K\sum_{i=1}^{n} p_i \log p_i$$

If, as is usually done, we take the logarithm to the base 2, then K=1.

(2) $$H(p_i) = -\sum_{i=1}^{n} p_i \log_2 p_i$$

The form of H happens to be the same as that of entropy as it is found in chemistry and statistical mechanics. This has often led to confusion especially in view of the fact that chemical entropy always carries dimensions. This is due to an incomplete transition from the anthropomorphism in which temperature is a scale factor which reflects "hot" and "cold". Temperature always appears in equations multiplied by the Boltzmann factor so that the product has the units of energy. Temperature really should be measured in energy units in the first place and it is increasingly popular to do so. Chemical entropy is a special case where we are considering the probability distribution of the energy states of the system under discussion. If we consider entropy as a primitive mathematical concept, it is clear that we may calculate the entropy of any probability distribution. The entropy of the probability distribution p_i in information theory provides a measure of the choice available to the sender and the uncertainty of the receiver as to which symbol will be received. Entropy is therefore what we mean by information. An interesting detailed discussion of entropy as a basic concept in statistical mechanics has been given (8). The reader may find it amusing to calculate the entropy associated with the casting of dice, the hands of cards, or an opinion poll.

The point which has often been misunderstood (9) and which we wish to stress is that we must be careful of confusing the entropy calculated from one probability distribution with that calculated from another. That is, entropy is not a quantity of itself but rather it is a property of a probability distribution. The chemical entropy of the probability distribution of the molecular or electronic energy exitation levels has nothing to do with the genetic entropy of the informational molecule.

A second source of misunderstanding and false expectations has been the notion that a theory of meaning could be constructed based on information theory. Such a theory will require additional principles to assign meaning to a series of symbols. Consider, for example, the following sentence with contains several series of symbols which are words in English, French and German: "O singe fort!" This sentence is nonsense in English and has rather different meanings in French and German. In biology the meaning of a series of symbols, if any, will pertain to specificity which is relevant to the cellular biochemistry. The reader may wish to be referred to a contrary opinion by WEAVER (20).

The first useful role that H can play in biology is that of an unambiguous measure of diversity. Suppose, for example, we have an ecosystem (16) and we wish to know how diversity varies with the season or with some other variable. We simply establish the population density in the species present and apply equation 2. Variations in H will then reflect the average change in diversity. One may also calculate the entropy for the subsystems as well.

Suppose, again, we wish to establish a measure of diversity for a class of protein. A large diversity will correspond to a large entropy and will reflect a greater complication. Hemoglobin, for example, has a large entropy and collagen a small one. Such a measure may be very useful in studies of protein evolution.

It is noted immediately that the information content of each amino acid selection is

$$H(y) = -\sum_{i=1}^{20} P_i \log_2 P_i$$

$$= +20 \frac{1}{20} \log_2 20$$

(3) $\qquad = 4.32$ bits/amino acid

The information content of the DNA base pair is 2 bits/base pair. This means that it is not possible to transmit genetic information for the most general protein unless the DNA codon contains on the average at least 2.16 base pairs. Therefore it is reasonable to expect that some of the codons will be triplets. Actually, of course, they are all triplets although in eight cases the third position is arbitrary and therefore these eight are essentially a binary code. The collection of sixty-four codon triplets has three functions:

a) Information storage and transmission
b) Punctuation
c) Redundance for error correction

We recall the fact that there is a transmission of specificity not only from one kind of molecule through several others but especially that this specificity is recorded in a four letter alphabet, transferred in another four letter alphabet and finally received in an alphabet of twenty-three letters, including punctuation. This indicates that some quantity is conserved in all these transcriptions and translations. The composition property shows that H has this required conservation property. A one-to-one code is the simplest code. It is so simple that it may not be clear to everyone that it is a code. Some workers call it a template and let it go as that. We should recognize that template action is a primitive one-to-one code and that a quantity of abstract character is being transferred. Eq. (2) is sometimes called the representation theorem.

The well known route of protein synthesis information which passes from the genome to the formation of the protein molecule is clearly an information recording and communication system replete with codes, redundancy and transcription errors, etc. as are all such systems. SHANNON's channel capacity theorem, which we will now discuss, therefore applies.

If we deal with the errors themselves, their generation in the genome, and final recording in the protein molecule, it is easy to be led to believe that information transmission is reduced by any errors whatever. That this plausible, but naive, assumption is not true is a second great advance in information theory. In fact, codes may be constructed such that a message may be transmitted with an arbitrarily small frequency of errors at any noise level up to the point where the output is completely random. The rate of information transmission is reduced but the message can still be transmitted.

In biology it is believed that the coding of information is universal and can't be changed. Nevertheless, the mathematical existence of other codes enables us to determine quantitatively the properties of the biological code and its ability to transmit the genetic specificity with the required faithfulness.

The capacity of the information source is calculated by equation (2). The information lost through noise is found by a conditional entropy function. Suppose we receive at the site of protein synthesis the symbols y_j when the symbols x_i are sent. There will be a probability distribution $p_j(i)$ which will reflect any uncertainty in the reception. (Read this as: the probability of x_i when y_j is known.) We also have a probability distribution of the paired events $p(i,j)$. SHANNON shows that the average uncertainty in x_i when y_j is known is

(4) $$H_y(x) = -\sum_{i,j} p(i,j) \log_2 p_j(i)$$

Since
(5) $$p(i,j) = p(j) p_j(i)$$

we have
(6) $$H_y(x) = -\sum_{i,j} p(j) p_j(i) \log_2 p_j(i)$$

The rate of information transmission is given as follows:

(7) $$R = H(x) - H_y(x)$$

There are two transitions probability matrices $p_j(i)$ and $p_i(j)$. They are related according to BAYES' theorem as follows:

(8) $$p_j(i) = \frac{p(i) p_i(j)}{p(j)}$$

Since information flows from x_i to the y_j, it is more convenient to put eq. 7 in that form. Let us substitute from eq. 8 in eq. 7 and we find that:

(9) $$R = H(x) - H_x(y) + \sum_{i,j} p(i) p_i(j) \log_2 \frac{p(i)}{p(j)}$$

In the genetic information application the first term $H(x)$ is the channel capacity of the DNA-RNA, $H_x(y)$ is the equivocation introduced by errors of recording and transition and the last term reflects the change in channel capacity at the t-RNA protein interface due to the gene code. This term vanishes if i and j assume the same values, i.e. 1, 2, 3, 4 and if all $p(i)$ and all $p(j)$ are equal. This term is easy to evaluate by choosing values for $p_i(j)$, as reflected by the gene code, in Table I.

The maximum value of $H(x)$ occurs when all codons are equally probable and all $p(i) = 1/64$. This means that the DNA nucleotides will be present in equal amounts, a condition which is found to be nearly the case in E.coli. Evaluation of the third term in equation 9 then is found to be 1.71 bits. In the noise free case, $H_x(y) = 0$, $R = 4.29$ bits per amino acid. This is slightly less than largest possible value for $H(y)$ obtained when all $p(j) = 1/23$, namely, 4.52 bits/amino acid. This is to be expected since from

(10) $$p(j) = \sum_i p(i) p_i(j)$$

Table I. The Transcription Probability Matrix Elements $p_i(j)$ of the Genetic Code

	1	2	3	4	5	6	7	8	9	10	11	12	13	14	15	16	17	18	19	20	21	22	23	24	25	26	27	28	29	30	31	32	33	34	35	i
x_i	U	U	U	C	C	C	U	U	U	U	A	A	C	C	C	C	A	A	G	G	G	C	C	G	G	G	C	C	C	C	A	A	A	A	G	CODON
y_i	A	G	U	A	G	U	A	U	C	G	A	C	U	A	G	U	A	G	U	C	A	U	G	U	C	A	G	U	C	A	U	C	A	G	U	CODON
																																				CODON
1 LEU	1	1	1	1	1	1	1	0	0	0	0	0	0	0	0	0	0	0	0	0	0	0	0	0	0	0	0	0	0	0	0	0	0	0	0	
2 SER	0	0	0	0	0	0	0	1	1	1	0	0	0	0	0	0	0	0	0	0	0	0	0	0	0	0	0	0	0	0	0	0	0	0	0	
3 ARG	0	0	0	0	0	0	0	0	0	0	1	1	0	0	0	0	0	0	0	0	0	0	0	0	0	0	0	0	0	0	0	0	0	0	0	
4 ALA	0	0	0	0	0	0	0	0	0	0	0	0	1	0	0	0	0	0	0	0	0	0	0	0	0	0	0	0	0	0	0	0	0	0	0	
5 VAL	0	0	0	0	0	0	0	0	0	0	0	0	0	1	1	1	1	0	0	0	0	0	0	0	0	0	0	0	0	0	0	0	0	0	0	
6 PRO	0	0	0	0	0	0	0	0	0	0	0	0	0	0	0	0	0	1	0	0	0	0	0	0	0	0	0	0	0	0	0	0	0	0	0	
7 THR	0	0	0	0	0	0	0	0	0	0	0	0	0	0	0	0	0	0	1	1	1	0	0	0	0	0	0	0	0	0	0	0	0	0	0	
8 GLY	0	0	0	0	0	0	0	0	0	0	0	0	0	0	0	0	0	0	0	0	0	1	0	0	0	0	0	0	0	0	0	0	0	0	0	
9 ILEU	0	0	0	0	0	0	0	0	0	0	0	0	0	0	0	0	0	0	0	0	0	0	1	0	0	0	0	0	0	0	0	0	0	0	0	
10 TYR	0	0	0	0	0	0	0	0	0	0	0	0	0	0	0	0	0	0	0	0	0	0	0	1	0	0	0	0	0	0	0	0	0	0	0	
11 HIS	0	0	0	0	0	0	0	0	0	0	0	0	0	0	0	0	0	0	0	0	0	0	0	0	1	0	0	0	0	0	0	0	0	0	0	
12 GLN	0	0	0	0	0	0	0	0	0	0	0	0	0	0	0	0	0	0	0	0	0	0	0	0	0	1	0	0	0	0	0	0	0	0	0	
13 ASN	0	0	0	0	0	0	0	0	0	0	0	0	0	0	0	0	0	0	0	0	0	0	0	0	0	0	1	0	0	0	0	0	0	0	0	
14 LYS	0	0	0	0	0	0	0	0	0	0	0	0	0	0	0	0	0	0	0	0	0	0	0	0	0	0	0	1	0	0	0	0	0	0	0	
15 ASP	0	0	0	0	0	0	0	0	0	0	0	0	0	0	0	0	0	0	0	0	0	0	0	0	0	0	0	0	1	0	0	0	0	0	0	
16 GLU	0	0	0	0	0	0	0	0	0	0	0	0	0	0	0	0	0	0	0	0	0	0	0	0	0	0	0	0	0	1	0	0	0	0	0	
17 CYS	0	0	0	0	0	0	0	0	0	0	0	0	0	0	0	0	0	0	0	0	0	0	0	0	0	0	0	0	0	0	1	0	0	0	0	
18 PHE	0	0	0	0	0	0	0	0	0	0	0	0	0	0	0	0	0	0	0	0	0	0	0	0	0	0	0	0	0	0	0	1	0	0	0	
19 TRY	0	0	0	0	0	0	0	0	0	0	0	0	0	0	0	0	0	0	0	0	0	0	0	0	0	0	0	0	0	0	0	0	1	0	0	
20 MET	0	0	0	0	0	0	0	0	0	0	0	0	0	0	0	0	0	0	0	0	0	0	0	0	0	0	0	0	0	0	0	0	0	1	0	
21 P	0	0	0	0	0	0	0	0	0	0	0	0	0	0	0	0	0	0	0	0	0	0	0	0	0	0	0	0	0	0	0	0	0	0	0	
+22 P	0	0	0	0	0	0	0	0	0	0	0	0	0	0	0	0	0	0	0	0	0	0	0	0	0	0	0	0	0	0	0	0	0	0	0	
23 P	0	0	0	0	0	0	0	0	0	0	0	0	0	0	0	0	0	0	0	0	0	0	0	0	0	0	0	0	0	0	0	0	0	0	0	

+ Punctuation

All p(j) are not equal if all p(i) are equal. The full channel capacity H(y) = 4.52 bits can be obtained but H(x) < 6 bits/codon because all p(i) are then not equal.

The redundance is defined as

(11) $$1 - \frac{R}{H(x)}$$

and in the above example is 28.5%.

In addition to the basic redundance of the code there is also redundance in the amino acid sequence. This is reflected in the substitutions which are found in homologous proteins in organisms which may have diverged from a common ancestor many million years ago. This redundance depends on specific cases and in general only average values can be given. This subject has been reviewed (12); it has been found that many substitutions are selectively neutral or make minor changes in protein function. This is not generally agreed upon (1), but for the purposes of this paper we will accept it.

There are eight amino acids in which the third position is arbitrary and which are, therefore, essentially coded by a doublet code. It is easily seen from eq. (2) that if at least five amino acids were completely arbitrary as far as protein specificity is concerned, or were not coded at all, then a doublet code could be constructed. This is a hint (10) that a doublet code may exist and was perhaps used at some early time in the history of life. It has been found (5) (13) that fifteen amino acids can be formed from ammonia, methane and water if subjected to an energy source. Several of the present 20 amino acids, namely methionine, trytophane, glutamine, asparagine and tyrosine are formed from simpler compounds by complex enzymic reactions. They may have been formed by some simple biotic or pre-biotic forms which used doublet code (13). Perhaps such forms contributed more complicated molecules to the pre-biotic nutritive soup. Such a code has no redundance and therefore no provisions for protection from errors. The modern code has the advantage of greater range so that it can code more amino acids, give better punctuation and provides redundance. For these and other reasons the forms, if they existed, could be expected to be replaced by the first organism to utilize the modern code.

III. APPLICATIONS TO BIOGENESIS AND EVOLUTION

One of the great principles on which modern biology is based is the concept of evolution by the force of natural or sexual selection on adaptive genes which are believed to arise by random mutation. We are now at a point in our discussion where we can apply information theory and molecular biology to the problem of biogenesis and evolution. These are in fact separate although related problems. In the first place we must show how life can arise from a primeval sea of complicated organic molecules (4) (14). That is, we must show how informational molecules carrying a primitive genetic message can arise by random substitutions so that the evolutionary process can act (17). In forming a theory of evolution we may start with a primitive informational molecule but we must then show how it can reproduce itself and how the information content is increased under selection pressure. Furthermore, our theory must give reasonably quantitative results or it is likely to prove the opposite of what we intend and we will be hoist by our own petard (21).

We have shown previously that a unique order of nucleotides in DNA is unstable (25). Non lethal mutations will soon produce viable variations in the order. There is also some arbitrariness in the amino acid sequence as homologous protein prove. Furthermore, each organism in the ensemble of organisms will carry a load of genetic noise. We must therefore consider organisms as members of an ensemble and the genome as an ensemble of genetic messages containing some genetic noise.

In using the term "load of genetic noise" we include the genetic load (26) of H. J. MULLER and those portions of the genome which carry no specificity or which carry a very large number of repetitions. It is important that this concept be operational. That is, we must be able to test at least by a Gedanken experiment whether or not a given series of DNA codons represents noise or is part of the ensemble of genetic messages. This is done by standard genetic methods. We suppose a frame shift mutation and ask whether any loss in capability can be found. If such a loss can be found the series contained information if not it contained genetic noise.

We have proposed a theory of aging, radiation damage and thermal killing based on the accumulation of genetic noise (24). The same idea has been suggested later from time to time by others (3) (9) (15). The same notion based on pre-code genetic ideas is embodied in the somatic mutation theory of aging and radiation damage. ORGEL (15) attempted to deal with the nucleotide errors themselves and was forced to introduce ad hoc an "error catastrophe". This "error catastrophe" actually follows from SHANNON's channel capacity theorem, eq. (7) as we have shown (24, 25). ORGEL suggested a very useful experiment to introduce genetic noise, namely, to incorporate appropriate amino acid analogues in the growth medium. This was indeed carried out (6) (7); the reader is referred to the cited articles for further discussion.

We are presently concerned with errors that tend to accumulate and will result in death both in the somatic line and in the germ line. Therefore even if we wish to understand just the continued existence of life irrespective of its origin and evolution, we must find a way in which the information content of the genome can be increased. That is, there must be a means of generating information just to correct for the accumulation of a load of genetic noise.

SHANNON has shown that there are 2^{NH} messages possible in a series of N symbols with information content H per symbol. The number of messages in a typical small protein of 300 amino acids is, in the noise free case, according to equation 9:

$$2^R = 2^{H(x) - B}$$

where B is third term on the right hand side of eq. 9. Therefore $2^{H(x)}$ is the total number of DNA messages and 2^B is the number of DNA messages which code a given amino message.

$$2^{300 \times 4.29} = 10^{387}$$

and

$$2^{300 \times 1.71} = 10^{154}$$

One can speculate on the amount of choice allowed by the specificity requirements of the protein message and calculate the number of the subensemble of active protein. QUASTLER (18) discussed the question of the expectation of biogenesis by random selection with a result similar to this.

In a recent paper was shown that the fantastic consequences reflected by numbers of this order of magnitude (19). In these calculations the redundance in the code or the redundance in the protein structure have not been taken into account. Even if we do so, the gene contains so much information that a given ensemble of genetic messages represents a vanishingly small fraction of the total ensemble population of all possible proteins. If we consider probability from the Bayesian point of view as a measure of degree of belief, we must rule out the notion that informational biomolecules as large as even the smallest modern ones arise de novo from the prebiotic soup.

It has often been suggested that some (up to 90%) of the DNA is not coded in relevant genetic messages. If we accept this idea tentatively, we may suggest that this portion of the genome is the site of the generation of genetic specificity which arises de novo by random processes and natural selection. Another source of genetic specificity is the gene which is already established in a section of otherwise unused DNA. We will discuss that presently.

Suppose we consider the primary specificity of a protein to be carried by a chain of a small number, m, of given amino acids. The expectation of this occurring by chance in a DNA chain of, say, 10^6 nucleotides is:

$$(\frac{1}{4})^{3m} \times \frac{10^{+6}}{3} \times 2^{-m} \times 1.71$$

The results are shown in Table II.

Table II. Expectation of a Pre-selected Series of m Amino Acids Coded by a Chain of 10^6 DNA Nucleotides

m	4	5	6	7	8
	1.8 x	8.4×10^{-7}	4×10^{-9}	1.9×10^{-11}	9×10^{-14}

The minimum number of amino acids required to specify an active site has been variously estimated to be at least seven. Such a site could arise by chance in a small but reasonable number of trials. Note that the expectation drops dramatically as m increases.

SPETNER (23) has calculated the amount of information which can be added to a nucleotide chain. He assumes a mutation rate of 10^{-8} per generation and finds for example that in $10^{12} - 10^{14}$ trials (i.e. 10^7 individuals x 10^7 generations) about 9.5 bits are generated in a chain of 1000 nucleotides if a fraction of about 10^{-3} of these chains have a selective value. This is enough information to add two specified amino acids in a protein chain of 333.

Only a few cases were considered, however. The subject of constructing and studying models of biogenesis and evolution has hardly begun. Such models can be extremely complicated and should incorporate all known genetic processes.

In a report on the rate of molecular evolution in hemoglobins of the lamprey, carp, bovine, rabbit and man (11), it was found that the rate of replacement is constant for each of the evolutionary lines mentioned above. This supports the hypothesis that the changes are controlled by the fortuitous appearance of a favorable combination. That is, the expectation time of incorporation of a favorable change is much less than the expectation time of appearance. The replacement rate is such that 4.29 bits are generated in a 140 member α hemoglobin chain in 3.9×10^6 years. This is about the rate calculated by SPETNER.

We may therefore see, at least tentatively, a way out of the dilemma discussed by QUASTLER

(18) and SALISBURY (19). In order to account for biogenesis it appears reasonable to consider means of information recording and transfer much more primitive that are found in modern organisms. We start with the assumption of a pre-biotic soup containing the required monomer units. It is reasonable to assume fairly high order polymers of random sequence. It is important to remember that the order of the monomer units is not determined by the potential energy pattern or other physico-chemical factors. The order of the monomer units will be selected at random.

Given that certain special series, seven or eight members in lenght, may have special catalysis or other properties we have shown that such series may be expected to arise from time to time. The most primitive code is one to one or pure template process (27). This occurs even in the modern protein synthesis and therefore can be involved here. We can imagine that the primordial soup may be considerably modified over a period of time as a result of this catalytic action. The environment may also have been modified by geophysical factors such as the loss of hydrogen from the atmosphere.

The next step we postulate is the invoking of a primitive doublet code which, as we have seen, can handle up to fifteen amino acids plus one codon for punctuation (10). Such a code can be expected to be quite inaccurate and to provide a poor information storage capability. The doublet code would, however, be the first real step from non-living to living matter since by our definition such matter would have an information content. Again we expect that this primitive life would change the chemical content of the nutritive solutions irreversibly over a period of time.

It is in this second stage of pre-modern biology that we suppose the first single organism appeared which used the modern code. Two facts, it seems to me, indicate that this was a rare single event and that the environment was again changed drastically so that only this new life could maintain itself. These two factors are the levo-rotary character of biomolecules and the universality of the biological code.

It seems clear that life should be possible with all dextro-rotary molecules. However, if the conditions from which modern life arose were changed in a time short compared to the expectation time for this life to appear again there would be no second chance. The code, of course, has often been cited as supporting the hypothesis of the single individual "Abraham" as the ancestor of all modern life.

It can be seen how we can avoid the dilemma discussed mentioned above engendered by the enormous variety of possible proteins. The generation and evolution of life must have occurred by a series of Markov processes. That is, a series of jumps between relatively stable configurations. There may therefore be a sort of network of specific proteins so that by a series of single substitutions or additions one may proceed along the chain. By the same token one would expect that in life which originated elsewhere other pathways would be chosen and that life forms would be quite different from those on earth. This chain of proteins has been illustrated by a word game (22). One may make the following series of English words by one point mutation at a time: WORD : WORE, GORE : GONE : GENE. The possibilities can be further illustrated by supposing that a replication is made such that WORD is repeated three times. We may then by means of point mutations, an insertion and an addition have the following series: WORD WORD WORD - WARD WARD SWORD - WARN WARP SWARD - WARN WASP SWARM - WARN WASP SWARMS - The final configuration is a message which might have a high survival value under some circumstances. So it may well be with proteins. As we explained above when a specific sequence which has biochemical specificity (meaning) appears and becomes part of the ensemble of genetic messages and thereby becomes fixed in the genome it reduces $H_y(x)$ and contributes to the information content of the genome. This process thus supplies information to make up or correct for that lost through noise and to contribute to that required for evolutionary change. It is therefore not necessary to introduce ad hoc the notion of "purposeful information" (9).

IV. CONCLUSION

We have sketched the way information theory can be used to consider the important current problems in biology. In many cases we have been able only to hint at a treatment. This will trouble some readers but we hope it will stimulate others to seek the answers of their own.

We must insist on this much: the genetic apparatus of the cell is a communication system. It is useless and idle to ignore this or to claim that the cell will not have all the properties of communication system. Some of our conclusions could be reached by combinatorial analysis but then the power of the representation theorem or the channel capacity theorem are not brought to bear.

As biology moves from a descriptive science to an exact one, mathematical theories will come more and more into play and into vogue. We have suggested that information theory can play a role similar to that of thermodynamics in physics, chemistry and engineering. Much can be done in those disciplines without thermodynamics. Much can be done in biology without information theory or without molecular biology for that matter. However, in each case a much clearer understanding is available in the realm of information theory or thermodynamics if these theories are used. In the indices of books on molecular biology and genetics one now finds listed the terminology of information theory. Perhaps it is high time that the full capability of the theory were recognized and applied. The study of the generation and destruction of genetic information can lead to new quantitative insight in biology.

References

1. CLARKE, B.: Science 168, 1009 (1970).
2. GAMOW, G.: Biol. Med., Kbh 22 (3), (1954).
3. GOLDMAN, S.: Perspective in Biol. and Med. 11, 12 (1968).
   GOLDMAN, S.: Perspective in Biol. and Med. 12, 638 (1969).
4. HALDANE, J.B.S.: Rationalist Annual 148 (1928).
5. HARADA, K., FOX, S.N. and ORO, J.: The Origin of Prebiological Systems and of Their Molecular Matrices, New York: Academic Press 1965.
6. HARRISON, B.J. and HOLLIDAY, R.: Nature 213, 990 (1967).
7. HOLLIDAY, R.: Nature 221, 1224 (1969).
8. JAYNES, E.T.: Phys. Rev. 106, 620 (1957); 108, 171 (1957).
9. JOHNSON, H.A.: Science 168, 1545 (1970).
10. JUKES, T.H.: Biochem. Biophys. Res. Comm. 19, 391 (1966).
11. KIMURA, M.: Proc. Nat. Acad. Sci. 63, 1181 (1969).
12. KING, J.L. and JUKES, T.H.: Nature 231, 114 (1971).
13. MILLER, S.L.: J. Am. Chem. Soc. 77, 2351 (1955).
14. OPARIN, A.: The Origin of Life, Academic Press, New York (1957).
15. ORGEL, L.E.: Proc. Nat. Acad. Sci. 49, 517-21 (1963); 67, 1476 (1970).
16. PIELOU, E.C.: An Introduction to Mathematical Ecology, Wiley-Interscience, New York (1969).
17. PONNAMPERUMA, C. and GABEL, N.W.: Space Life Sciences 1, 64 (1968).
18. QUASTLER, H.: The Emergence of Biological Organization, Yale University Press, New Haven (1964).
19. SALISBURY, F.B.: Nature 224, 342 (1969).

20. SHANNON, C.E. and WEAVER, W.: The Mathematical Theory of Communication, The University of Illinois Press (1949).
21. SIMPSON, G.C.: Science 143, 769 (1964).
22. SMITH, J.M.: Nature 225, 563 (1970).
23. SPETNER, L.M.: J. Theoret, Biol. 7, 412-419 (1964); IEEE Trans. Info. Theor. 11, 3 (1968); Nature 226, 948 (1970).
24. YOCKEY, H.P.: Radiation Research 5, 146 (1956).
25. YOCKEY, H.P., PLATZMAN, R.L. and QUASTLER, H.: (Eds), Symposium on Information Theory in Biology, Pergamon (1958) p. 50 et seq.
26. WALLACE, B.: Genetic Load in Biological and Conceptual Aspects, Prentice-Hall, New York (1970).
27. WOESE, C.R.: Proc. Nat. Acad. Sci. 59, 110 (1968).

DISCUSSION

BRUNNGRABER:
FONG's thesis (this symposium) rests on the assumption or definition that the first sign of life is the development of the self-replicating molecule. Subsequent biological events involve modification of the kinds of information transmitted by the progeny of the primordial self-replicating molecule. This assumption is also implicit in YOCKEY's presentation. This assumption serves the purpose of permitting us to avoid the intellectual difficulties involved in comprehending the events we call <u>biogenesis</u>. In view of our ignorance of events leading to biogenesis, it seems to me that alternative definitions may be equally valid. For example, a nonreproducing molecule may have served as a template or matrix for the formation of the first molecules which subsequently acquired the capacity for self-duplication. In this case, the type of information carried by the primordial multiplying molecule would have been conditioned by the nature of the information transmitted by the non-reproducing template. Possible mechanisms have been discussed by other authors (C. PONNAMPERUMA, L. CAREN and N. GABEL, in: "Cell Differentiation"; edited by O.A. SCHJEIDE and J. de VELLES, pp. 15 - 30, Van Nostrand and Reinhold Co, New York, 1970). The process of biogenesis may have had its own developmental history for a long period of time prior to the development of the self-replicating information-transmitting biomolecule.

YOCKEY:
The main thrust of my paper is that to understand biogenesis and evolution as well as aging and other problems in biology one must understand the creation, storage, transfer, and degradation of genetic information. This will help in solving not avoiding the intellectual difficulties involved in comprehending biogenesis. I certainly agree that there may well have been a developmental history prior to the appearance of the most primitive modern form of life.

WALTER:
Aren't the calculations on page 12 and the statement that the DNA codon must contain more than two base pairs based on the assumption that the codon is not three (or larger) dimensional?

YOCKEY:
No. This calculation is based on an abstraction, namely, the information content of the choice.

FONG:
I am afraid that once the genetic code began as a doublet code it cannot evolve into a triplet code because any such change would be a lethal mutation. This point and many other characteristics of the genetic codes are discussed in a theory of the nature, origin and evolution of the genetic codes I have developed (Biophys. J. Soc. Abst. 11, 261a (1971)).

In this theory the triplet nature of the code is determined by the molecular size of t-RNA. In the early stage of evolution, however, the third letter of the triplet is a dummy space and so the code is effectively a doublet code from the information point of view. The degeneracy of the genetic codes can be explained by an evolution process based on mutation from the less probable to the more probable, not on the advantages from the information theory point of view. - The important role played by information theory in biology does not imply that biological systems are pre-designed based on information theory considerations. This point is also related to the problem of QUASTLER and SALISBURY. We should not pre-conceive a set of biochemical processes and a set of necessary macromolecules (such as insulin) and ask the evolution process to bring them about. The probability of realizing these by random chance is practically zero. On the other hand if we adopt the view on origin of life advocated in my paper in this volume, this problem disappears automatically. To illustrate by an analogy: If a man is to look for a wife that is exactly the same as M.M. or B.B. the chance of succeeding is nearly zero. Yet most men get married -- the function of marriage may be fulfilled by one partner or another and in the "evolutional" process someone will be found.

YOCKEY:

The paper does not say that the hypothetical pre-modern binary code evolved directly into the modern triplet code. This is not excluded but the main point is the speculation that this pre-modern life modified the environment to favor the emergence of the modern gene code. The evolutionary selection process proceeds, of course, from more probable to less probable, i.e. from less selective to more selective. - Those who did the brilliant work on which our knowledge of the code is based were annoyed and puzzled by "degeneracy" and "nonsense codes". This rather negative terminology reflects this attitude. However the code evolved and whatever the molecular structure considerations, the point is that the function of "degeneracy" is to provide error protection through redundance. The "nonsense codes" of course provide punctuation as is well known. - Leaving the wisdom of the choice of B.B. or M.M. for a wife aside, the apriori probability is about 3×10^{-10}. (M.M. is no longer in the sample space.) If we assume a prejudice factor of 10^{-3} to reflect contiguity and pecuniary factors we have 3×10^{-13}. This is fantastically greater than the probability of a particular protein in a 300 element chain. We may assume that 10^{12} of them have specificity and still find it incredible that any one of these could be emerge in one step like Aphrodite from the sea foam by purely random processes. Our experience justifies being far less sure that the sun will rise tomorrow or that a pot of water set on a cake of ice will not boil. For the reasons discussed in the paper is seems worthwhile to look for mechanisms whereby information may be generated and accumulated over a very long period of time.

BREMERMANN:

Mrs. GATLIN has computed various information theoretical quantities for DNA of different species (redundancy, etc.) and she has found differences for bacteria, mammals and viruses. She has various hypotheses about the significance of these differences and she has written several papers and a book about it. - N. GOEL, YCAS, Lacy KING and I have been computing actual codon frequencies in place of the apriori frequencies of Table I. These computations determine the frequencies from experimental data such as nearest neighbor frequencies, isostitches, etc. The computation involves the solution of non-linear equations in 64 variables which has required novel mathematical-numerical methods. A paper is in preparation. - M. EIGEN has developed a new theory of the chemical evolution of DNA and RNA polymers in a "primordial soup". His theory sheds light on how the first polynucleotides (and proteins) arose. - I (BREMERMANN) have drawn conclusions from the amount of information that is contained in the genetic system (IEEE Transact. 7, 200, 1963; Progr. Theor. Biol. 1, 59, 1967). The more DNA, the more complex the resulting organism can be. The more DNA, the slower an organism can evolve and adapt to changes in the environment. Viruses, bacteria and vertebrates constitute different compromises between these two opposing constraints.

YOCKEY:
In general these references (and others, kindly supplied by Professor BREMERMANN and Mrs. GATLIN) support the points made in my paper. Some of this material is still in press. The papers are recommended to the reader who wishes to learn more of the usefulness of information theory in biology. GATLIN, L.L.: J. Theoret.Biol. 10, 281 (1966), 18, 181 (1968) - GATLIN, L.L.: "Evolutionary Indices" Proc. 6th BerkeleySymp.on Math.Stat. Prob.(1971) (in press) - GATLIN, L.L.: "The Entropy Maximum of Protein" Math. Biosc. 13, 213 (1972) - JUKES,T.H. and GATLIN, L.L.: "Recent Studies Concerning the Coding Mechanism" Progress in Nucleic Acid Research, Vol. II, Academic Press, Inc., New York (1970) - REICHERT T.A. and WONG A.K.C. "Toward A Molecular Taxonomy" J.Mol. Evolution, Vol 1 (1971) (in press) - EIGEN, M.: Die Naturwissenschaften, 58, 465 (1971).

LOCKER:
Your paper clearly demonstrates how far information theory can contribute to an understanding of the origin of life and of evolution. In my opinion your remark that information is existing only in living and not in non-living entities is an expression, in other words, of the fact that living things display some properties resembling a subject-like self by means of which they "understand" the meaning of the information they receive. However, in as much as information is an abstract (or formal) measurable quantity it can merely be considered independently of its carrier or transmitter whereas information can "act" - provided it is able to act at all - only in connexion with such a carrier (into which it can also be stored). You very convincingly offer an explanation for the possibility of information increase. In this respect, it is not clear to me in which way this increase would constitute a compensation process for the accumulation of noise. A simple shift from noise to information - which would formally and quite naively appear as similar to the emergence of organization at the costs of pre- or un-organized entities - is of course excluded by SHANNON's theorem on the irreversibility of the information to noise transformation. Therefore, by necessity, the preexistence of a system comprising information as well as noise as distinct "parts" must be assumed. No process that supplies an information (i.e. meaning) can actually occur unless a system exists with reference to which the, e.g., randomly arisen informative configuration (sequence etc.) of molecules, makes sense. I can completely agree with the idea that an increase of information occurring at random was favored by propitious environmental conditions, or even maintained for a while by such conditions. But can this idea also be upheld for the origin of information? It seems to me - venturing an idea being in vague analogy to GOEDEL's theorem - that the origin of information cannot be explained by information theory itself. This suspicion is obviously shared by several workers and possibly the reason why on the one hand the necessity to define a qualitative information is frequently postulated and why on the other hand arguments have been put forward claiming a theory of algorithms (or programs or instructions) which give the directions according to which information can be transmitted. The comparison has been made with vocabulary and grammar, respectively. Such a dynamic theory should enable us to comprehend how information really can act thus exceeding its so-to-speak passive role of a quantity subsumed as invariant when at best it is preserved and which at worst is destroyed or annihilated. In my conviction even such a theory of algorithms could not explain wherefrom the algorithms, possibly detectable in living organisms and thus directing, e.g., information transmission, descend.

YOCKEY:
This comment considers the crux of the problem, namely, that living matter seems to flout the second law of thermodynamics. We can understand that information would gradually be lost through noise. But how can it be created or even maintained? Unlike thermodynamics or electrical communication there is in living matter a value function attached to success which results in successful genetic messages becoming more numerous. Let us suppose in a certain city all safes had three combination locks and were owned by an excentric billionaire who, for his convenience, set them all at the same combination. Burglars might attempt to open

these locks by trying all possible combinations. Sooner or later one would succed and then he could open all the safes. The burglar has thus gained information from his environment. If he transmits his code through a noisy channel to a confederate and one number is incorrectly received, the confederate nevertheless has a much easier time to finding the correction. Soon this combination will be common knowledge among all the burglar's confederates. Thus, this information will have high survival value and thus will be stored and reproduced.

The value of a given message to the receiver is indeed outside of information theory and must be established by other means. This was illustrated in the comment about meaning in the paper. The theoretical biologist, like a carpenter, needs a full kit of tools. I advocate that information theory be one of these but not the only one.

On the Origin of Information in Biological Systems and in Bioids

P. Decker, A. Speidel, and W. Nicolai

Abstract

Some general principles of evolution and accumulation of information can be discussed independently of the special case of terrestrial life using the concepts of bioids, open systems which can exist in several steady states. Extending classical information theory, three "kinds" of information can be discussed: I_1, information representing the chemical ("chemical information") and physical constraints, that bioids must be capable to evolve and to make operative in some environment, is shown to depend on the observer and his deficiency in the knowledge of natural laws; I_2, stochastic information depending on the part of operable systems which actually arose during evolution. It is postulated that "useful" information cannot accumulate in I_2. I_3, environmental information depending on the part of systems that survived selection, is interpreted as a mapping of sets of environmental conditions into sets of systems by the process of selection. Fitness of the environment can be shown to result as a consequence of "common information". Individuation can modify the control of information increase by supplying negentropy. Some possibly reproducible first steps of prenucleoprotid bioid evolution, are selected for maximal information increase: Autocatalytic formation of sugars from formaldehyd, coupled to light energy through photosensitized oxidation of methane by water, and individuation by production of colloid or insoluble matter, associated with division or budding caused by osmotic pressure.

References

DECKER, P.: Nature, in press (1972); J. Molec. Evol., in press (1972).
DECKER, P., SPEIDEL, A.: Z. Naturforsch. 27 b, 257 (1972).

On the Evolutionary Origin of Life and the Definition and Nature of Organism: Relational Redundancies

J. R. Hamann and L. M. Bianchi[+]

Abstract

The relational systems formalism can be applied to the evolutionary origin of life and the definition and nature of "organism". In particular an "organism" (biosystem) is defined as a stable aggregate of biosystem components, the latter in turn being specified in terms of the existential realization of the necessary and (possibly) sufficient conditions for the evolutionary origin of life. The difference between biosystems and physico-chemical systems can further be explicated in terms of the respective system dynamics, with special attention to the probabilistic mechanics and the prior probabilistic decision process. It can be noted that a generalized statistical mechanics and thermodynamics are subsumed by these latter. The resolution of the problem of irreversibility can be discussed along with general approaches to the analysis of stability and instability (or the onset thereof) and the applicability of control systems as models for biosystems. On this basis, molecular (e.g. metabolism, chemical learning), cellular (e.g. membrane depolarization; probabilistic neuromechanics), organismic (psychosystems; value and optimization; language and meaning) and population (VOLTERRA - LOTKA - KERNER systems) levels in biosystem hierarchies can be discussed.

References

BIANCHI, L.M. and HAMANN, J.R.: J. Theor. Biol. $\underline{28}$, 498 (1970).
HAMANN, J.R. and BIANCHI, L.M.: J. Theor. Biol. $\underline{28}$, 175 (1970).

DISCUSSION

FONG:
Because of lack of explanation of many of the special concepts introduced it is difficult to understand some of the discussions without a previous knowledge of this line of development.

I do feel that the concepts of relation and information can be unified (as I did in my paper in this volume). Relation is just one kind of information. The basic principles are the same.

ROSEN:
Without going into the specifics of the authors' developments it should be pointed out that their use of the term "relational" seems rather different from that used by RASHEVSKY in his discussions of "relational biology" starting in 1954. As I see it, RASHEVSKY's use of the term was in connection with extracting general principles common to all organisms on a particular level of organization; a corresponding statement in physics might be to note that the equations of motion of particular optical, mechanical or electro-dynamical systems might be described by a common variational principle. The present authors use the term mainly for describing how we might be able to pass effectively from a theory at one level of organization to corresponding theories at other levels of organization. It seems to me that an unqualified use of the same

---

[+] This work was supported in part by Grant NGR-33-015-016 from NASA

term in two such different contexts might generate a certain amount of confusion, although I am at a loss to make a concrete suggestion as to how to alleviate it.

HAMANN, and BIANCHI:
Final remark not received

On the Dynamics and Trajectories of Evolution Processes

H. J. Bremermann

Abstract

René THOM has described a general theory of morphogenesis. He points out the importance of attractors of systems trajectories and of catastrophes (generalized bifurcation points). The present paper suggests to replace the dynamic law in the form of differential equations by probabilistic laws in order to describe evolution processes. The main features of THOM's theory remain valid, in particular attractor phenomena can be interpreted as corresponding to biological species. For the case where a function exists that corresponds to the potential function of a gradient field the author has observed attractor phenomena in simulations of evolution processes. Finally, connections with optimization algorithms are pointed out.

René THOM has emphasized the importance of topology for morphogenesis. His concept of morphogenesis covers not only embryology but includes the growth of forms in the most general way ((15), p. 152).

Morphogenesis, the growth of forms, is a property of <u>dynamical systems</u>. A dynamical system is a collection of states and a <u>mechanism of transition between states.</u> In many (but not all) cases the states form a manifold. <u>Dynamical laws</u> describe the mechanism of state transition. In many (but not all) cases the dynamical laws take the form of differential equations that express derivatives of the state variables in terms of a vector or gradient field on the state space. The notion of dynamical system is so general that it covers embryogenesis, biological evolution, enzymatic, metabolic and general chemical systems as well as mechanics, quantum mechanics and electrodynamics.

For example, a chemical system can be described as follows: Consider chemical substances $s_1, s_2, \ldots s_k$ of concentrations $c_1, c_2, \ldots c_k$. Because of chemical reactions between the substances, the concentrations vary in time according to some rate law

$$\frac{dc_i}{dt} = X_i(c_1, \ldots c_k), \quad i = 1, \ldots k.$$

For example, the X_i could be second or third order functions in the c_i determined by the mass action law. The X_i form a vector field on a subset of Euclidean space. The development of the system in time is described by the point $\vec{c}(t) = (c_1(t), \ldots c_k(t))$, that moves along a trajectory, in $\vec{c}$ space.

In many cases the trajectories will tend asymptotically to a limit state c_0 but other cases are also possible: limit cycles, manifolds or even pointsets with less regular structure. A connected set of limit points is called an <u>attractor</u> of the system. This definition applies not only to chemical systems but to systems in general. Given an attractor (A), the set of trajectories that tend to (A) is called the <u>basin</u> of (A). A system may have several attractors and it is possible that attractors have common boundary points and that they interpenetrate in topologically complicated ways even though the system is structurally stable (small perturbations do not change the topology). THOM calls points such that an infinitesimal perturbation can result in different trajectories that belong to different basins <u>catastrophic points</u>. Catastrophic points can exist on structurally stable attractors.

In THOM's view the existence of catastrophic points makes possible the morphogenesis of a great variety of different forms (such as in embryogenesis) on the basis of a single <u>deterministic</u> process ".... deterministic systems may exhibit in a 'structurally stable way' a complete indeterminacy in the qualitative prediction of the final outcome of its evolution." ((16), p. 317) In a practical sense THOM hopes that "Our dynamical schemes - with the ideas of attractors, bifurcation, catastrophes ... - provide us with a very powerful tool to reconstruct the dynamical origin of any morphological process." ((16), p. 334) In embryogenesis the underlying dynamical process is unknown. If THOM's concepts would be the key to finding it, they would solve one of the big mysteries of biology.

Before we turn to the main subject of this paper, biological evolution, let us consider further examples: A mechanical system can be described by means of the Hamiltonian equations

$$\frac{\partial p_k}{\partial t} = -\frac{\partial H(\vec{p}, \vec{q})}{\partial q_k}$$

$$\frac{\partial q_k}{\partial t} = \frac{\partial H(\vec{p}, \vec{q})}{\partial p_k},$$

where $\vec{p}$ and $\vec{q}$ are local coordinates on a manifold and H is the Hamiltonian function. Analogously electromagnetic systems can be described by electro-magnetic vector fields that are governed by MAXWELL's equations as dynamics.

In many cases (under rather general conditions) the vector field is a gradient field, generated by a potential function V. The critical points, that is the points where grad V vanishes, are equilibrium points of the system. THOM lists some of the most basic singularities of potential functions in one and two dimensions, and he gives spatial and temporal interpretations of the associated catastrophic points. THOM is pessimistic about experimental verification, however: "Ces modèles sont-ils susceptibles d'un contrôle expérimental? Il ne faut, au moins pour le moment, repondre à cette question par la négative." ((16) and (15), p. 160)

I have observed phenomena which while they do not constitute an experimental verification of THOM's models of elementary catastrophes, are closely related. They involve potential functions and their singularities and the phenomenon of attractors and breaking loose of trajectories. Ironically, in my case the underlying dynamic process is not deterministic while THOM is much concerned with showing how apparent indeterminacies can arise out of deterministic processes. He even views in this way biological evolution which is commonly understood as being based on random mutations. "That mutations have a character which is strictly random (if such an expression has sense) is one of the dogmas of present-day biology. It seems to me however, that this dogma contradicts the physical principle of action and reaction; ..." ((14), p. 38) THOM goes on to give reasons for his preference of a deterministic process underlying evolution. ((14), pp. 38 - 41)

Philosophy aside, one gets a simpler if phenomenological description of evolution if one settles for a probabilistic dynamics of mutation and recombination. Our description is very similar to that of the deterministic systems that we have been concerned with so far. There is a collection of states with a structure and a dynamics of transition between states. The dynamics, however, is probabilistic.

Given several species in a fixed physical environment. The species interact with each other and with the environment. The reproductive success of each individual depends upon its genotype, upon the resources of the environment, and upon the other organisms -- characterized by their genotypes -- that compete for their resources. A genotype may be characterized by gene variables $g_1, \ldots g_n$ each of which can range over a set of discrete states called alleles. Rather than work with the growth of species, we follow the propagation of individual genotypes. A species is a set of genotypes that are nearly identical in the sense that they differ in only a few gene variables.

We number all genotypes consecutively from 1 to N. In general N is very large. If there are k alleles per gene, then $N = k^n$. N, in general, is greater than the number of particles in the universe. Hence, many of the X_i are zero.

The growth rate $\frac{dX_i}{dt}$ of genotype X_i can be written as the sum of the two terms $f_i + g_i$ where f_i depends only upon the concentration of the genotypes $X_1, \ldots X_N$ and upon environmental factors such as the concentration of nutrients, precursors, and energy, while

$$g_i = \sum_{j \neq i} M_{ij} X_j - X_i \sum_{j \neq i} M_{ji} ,$$

where M_{ij} is the probability that genotype X_i mutates into genotype X_j. The probabilities M_{ij} describe the mutation dynamics, while the f_i describe the competitive "success" of genotype X_i. (A theory along similar lines was communicated to the author by M. EIGEN (12)).

Instead of numbering the genotypes $X_1, \ldots X_N$ we can also write $X(g_1, \ldots g_n)$, where $g_1, \ldots g_n$ are the gene variables. If the f_i vary slowly with respect to small changes in the concentrations of the $X_1, \ldots X_N$, then we can introduce a function $V(g_1, \ldots g_n)$ that is analogous to a potential function in deterministic vector fields on a manifold. In genetics it is known as a fitness function. The dynamics of the process is such that $X(g + \delta g)$ increases if V is larger at $g + \delta g$ than at g. It decreases if V is smaller at $g + \delta g$.

The competitive success of genotype depends here only upon V and not upon the concentrations of other genotypes. It holds locally if we try out new genotypes against an environment where the concentrations of other genotypes remain approximately constant.

The dynamics of the $\delta g_1, \ldots \delta g_n$ depends upon mutation and mating. In the simplest case each δg_i occurs with a fixed mutation probability p, independent of the other δg_j. In this case the probability that a particular pair mutates is p^2, the probability for a particular triplet mutating is p^3, etc.

The trajectories of such a process (for a quadratic "potential" function V) were investigated by the author several years ago ((2, 7 ... 10)). At that time, THOM's theory did not yet exist. The author was very surprised when he observed that the trajectories of the process would not proceed smoothly to the (unique local and global) maximum of V but instead would stagnate at points other than the maximum. These points were stable except for the rare occurrence of certain multiplet mutations (simultaneous mutation of several genes). This phenomenon persisted no matter how he modified the particular details of the process (2). Not even mating helped much. Mating is the recombination of genotypes (within a species) to form new genotypes according to a rule of random recombination or according to some other mechanism.

The peculiarity of mutation dynamics is that the increments $\delta g_1, \ldots \delta g_n$ are not uniformly distributed but that the "coordinate axes" are biased! Simultaneous mutations of more than a few genes are rare. Thus most of the components of δg are zero. It turned out that the evolution process stagnated exactly at points where V increases for all δg with only a single (or a few) nonzero components. Since mating takes place only between closely related genotypes that differ by at most a few gene variables, it produces again δg with most components equal to zero and thus does not easily free a trajectory from its attractor. The points of V where all single $\Delta V = (g + \delta g_i) - V(g)$ are negative, are analogous to critical points, and those critical points that are not maximum correspond to saddle points. It is the latter that attract the evolution process and that make it stagnate away from the maximum (2). The very same kind of points are discussed by THOM as generating the elementary catastrophes for deterministic processes.

Evolution is an optimization process. It has a particular process dynamic that is hard to analyze because of the non-uniform generation of increments δg.

Abandoning evolution one can pose the following problem: given a function V of a number of real variables. Problem: design a dynamic process that will generate trajectories that approach the (absolute) maximum of V as rapidly as possible. This means in other words: Find the most efficient algorithm of (global) optimization. Inspite of a profuse literature (well over a thousand papers) this problem is far from solved. Most optimization algorithms proceed along the gradient of V. Unfortunately such algorithms will lead only to local optima and get these in a way that is not necessarily the fastest.

On the basis of his numerical experimentation with evolution and similar processes this author has developed an optimization process that proceeds in random directions rather than along the gradient and that determines the maximum along this direction globally.

Under suitable, rather general conditions this process does not stagnate indefinitely at local maxima. It will eventually get to the absolute maximum (or minimum). Local maxima, however, may attract the process for periods of time, the lenght of which depends upon the particular properties of the function V (6).

Optimal control also fits into this general framework. Here an objective function, defined on a space or manifold, and partially determined local dynamic rules are given, usually in the form of differential equations. Some variables or functions are to be determined by the <u>controller</u>. In the previously discussed problem there was complete freedom to design the dynamics of the process. In this case the local dynamics can only be modified by the choice of values for the control variables and functions. The goal is to control the local dynamics so that the resulting process trajectory reaches a desired point as quickly as possible or in some other optimal way (such as with minimum fuel consumption, etc.).

All these different systems and processes fit into a single conceptual framework. In this author's point of view the restriction to deterministic processes is very unnecessary. The theory of probabilistic processes and their resulting orbits should be studied in the same way as deterministic processes have been studied ever since H. POINCARÉ founded the discipline of <u>qualitative</u> mechanics+. The reason for a qualitative rather than quantitative approach to global trajectories lies in the fact that even the most simple cases are not amenable to explicit

---

+ (For a survey of the current state of the art in this field see (13)).

analytic solutions. (That the difficulties of explicit analytic analysis of the probabilistic case are equally formidable is apparent in (6)). If an explicit quantitative global analysis is intractable to analytical tools, it also remains, in many aspects, intractable to computers. Many problems are "transcomputational" (1) , though exploration by means of computational experimentation is, of course, a very valuable tool.

Conclusion and Conjecture

The description of systems by a state space and dynamic laws is so general that it comprises almost any conceivable physical, chemical, biological, ecological, ecomomic, or social system. Independent of the question whether the world is ultimately deterministic or not, probabilistic state transition laws can be valuable as phenomenological descriptions. Probabilistic dynamics, however, at least in the continuous case has hardly been studied at all.

Whatever the specific dynamics and structural details of the process, the capturing of trajectories by attractors seems to be a fairly common phenomenon, even if the dynamic laws are deliberately designed to avoid this as much as possible (as in optimization). While THOM speaks of <u>catastrophes</u> the breaking loose of a trajectory from an attractor (from a "hang-up") may also be viewed as <u>liberation or breakthrough</u> when the process is optimization or evolution.

Our conjecture is that in biological evolution the <u>species</u> correspond to states that have been captured by attractors. The latter correspond to the ecological niches. Species, thus, are not optimal in a global sense but only in a local sense. This point of view has been expressed by the author (2) (10), p.6) as a result of his simulation experiments. THOM comes to the same conclusion on the basis of his general conceptual homework of morphogenesis: "Thus evolution is nothing other than the propagation of an immense wave front across this paleogenic space ..., the motive phenomenon is probably the attraction of form, every proper form (I would say an "archetype" if the word did not have a finalistic connotation) aspires towards existence and attracts the wavefront of existing beings when that has reached neighboring topological forms; there is competition between the attractors, ... Of all the living forms theoretically possible only an infinite minority are touched by the wavefront and come through to existence " (14, p. 40 - 41). THOM speaks here of 'wavefronts' rather than points of trajectories. This depends upon the description of the process. In evolution it is convenient not to describe each possible combination of possible genotypes as a single point but to have a much lower dimensional genotype space plus a number that indicates how much of this particular genotype exists (we could measure the number of individuals or the amount of DNA). For most genotypes this number will be zero. In this description the individual state of the system is thus not a point but a hypersurface, and the theory becomes one of the propagation of probabilistic hypersurfaces.

While these conceptual frameworks are important and may be considered to exist in a Platonic sense the drawing of inferences is limited in two ways: Mathematical inferences are limited by the intrinsic undecidability of many questions (11) and computational exploration is limited by a fundamental (quantum- theoretical) limit of data processing (1), (2), (3), (5).

References

1. ASHBY, R.: In: OESTREICHER, H.L. and MOORE, D.R. (Eds.), Cybernetic Problems in Bionics, p.69, New York: Gordon & Breach 1968.
2. BREMERMANN, H.J.: In: YOVITS, M.C., GOLDSTEIN, G.D. and JACOBI, G.T. (Eds.), Self-Organizing Systems, p.93. Washington, D.C.: Spartan 1962.
3. BREMERMANN, H.J.: Progr. Theor. Biol. <u>1</u>, 59 (1967).
4. BREMERMANN, H.J.: In: OESTREICHER, H.L. and MOORE, D.R. (Eds.) Cybernetic

Problems in Bionics, p.597, New York, Gordon & Breach, 1968.
5. BREMERMANN, H.J.: Proc. Fifth Berkeley Symp. Math. Stat. a. Prob. $\underline{4}$, 15 (1968)
6. BREMERMANN, H.J.: Math. Biosc. $\underline{9}$, 1 (1970).
7. BREMERMANN, H.J. and SALAFF, S.: Experiments with Patterns of Evolution, Techn. Rept., Contr. Nonr $\underline{222}$ (85) & 3656 (08), Berkeley, 1963.
8. BREMERMANN, H.J. and ROGSON, M.: An Evolution-Type Search Method for Convex Sets, Techn. Rept., Contr. Nonr $\underline{222}$ (85) & 3656 (08), Berkeley, 1964.
9. BREMERMANN, H.J., ROGSON, M., and SALAFF, S.: In: MAXFIELD, M., CALLAHAN, A. and FOGEL, L.J. (Eds.), Biophysics and Cybernetic Systems, p. 157, Washington D.C. Spartan, 1965.
10. BREMERMANN, H.J., ROGSON, M., and SALAFF, S.: In: PATTEE, H.H., EDELSACK, E.A., FEIN, L. and CALLAHAN, A.B. (Eds.), Natural Automata and Useful Simulations, p.3, Washington, D.C., Spartan, 1966.
11. DAVIS, M.: The Undecidable, Basic Papers on Undecidable Propositions, Unsolvable Problems and Computable Functions, New York, Daven Press, Hewlett, 1965.
12. EIGEN, M.: Private Communication, August 1970 (now appeared in: Naturwiss. $\underline{58}$, 465 (1971)).
13. SMALE, S.: Bull. Am. Math. Soc. $\underline{73}$, 747 (1967).
14. THOM, R.: In: WADDINGTON, C.H. (Ed.), Towards a Theoretical Biology I: Prolegomena, p.32, Chicago, Aldine Publ. Co., 1968.
15. THOM, R.: as in (14), p. 152.
16. THOM, R.: Topology $\underline{8}$, 313 (1969).

DISCUSSION

ROSEN:
This most interesting paper suggests a number of comments and questions. I will only raise a few of these, suggested by my own work on dynamical systems in biology. - First, the dynamical formalism has even more flexibility than is perhaps indicated in the paper under considerations, or in the work of THOM. For instance, we typically consider only the system trajectories and their orientation (unless we should be so fortunate as to have a reasonable set of dynamical equations given to us). But the trajectories themselves are curves parameterized by time, and this parameterization is another "degree of freedom" of the system which is at our disposal. For instance, the simple two-dimensional system

(1) $\quad \dot{u} = -\lambda_1 u, \quad \dot{v} = \lambda_2 v$

has exactly the same trajectories as

(2) $\quad \dot{u} = -\lambda_1 u f(u,v), \quad \dot{v} = -\lambda_2 v f(u,v)$

where f (u, v) is an arbitrary function which does not vanish on the state space. Thus, the trajectories of (1) approach the unique attractor at a definite rate, which we can increase without bound by suitably choosing f(u, v). On the other hand, by slowing this rate down, we can keep a system near a catastrophic point for as long as we like. - Second, the mere fact that a system fails to be (structurally) stable to all possible perturbations does not mean it is without interest. If we know that the system is conditionally stable, then it can still model biological processes, with the proviso that, under "normal operating conditions", the system never sees a perturbation which would elicit the instability. This is a point frequently overlooked in analog simulations, which are frequently plagued by instabilities and therefore discarded; the point is that all the rheostats of an analog computer are equally easy to reset, while the corresponding parameters of the system being modeled may be exceedingly hard to perturb. Thus simple conditional stability becomes in fact a powerful tool for probing the circumstances in which a given biological system was actually designed to operate (e.g., there is no reason to expect global stability to evolve in circumstances when conditional stability will suffice). The same point arises with conservative (Hamiltonian) systems, which are typically not structurally stable, but are conditionally structurally stable. - Finally, I would point out that the species space described in the paper under consideration is an example of the situation I discussed in my own paper; namely, a dynamical system whose states do not correspond to the states of a single real system in real time.

FONG:
The capturing of trajectories by attractors discussed in this paper corresponds to the principle of perfection in my paper and the liberation or breakthrough mentioned here may be achieved by the principle of imperfection in my paper. The mathematical expression of the capture of trajectory is the maximizing of the fitness function which is analogous to the potential energy function. An additional mathematical innovation is needed to incorporate the act of liberation. Perhaps this may be accomplished by introducing a function somewhat analogous to the kinetic energy function together with a statistical treatment of the trajectories. In this way liberation or breakthrough may be expected to become a regular part (not merely a singular accident) of the evolution process.

LOCKER:
Your conjecture that a species is not optimal in a global sense but only in a local one seems 1. to agree with THOM's statement that the evolutionary development proceeds in conformity with a purely local determinism and 2. to be in accord with the general and widespread belief among scholars that the motive processes of evolution are those of micro-evolution and not of macro-evolution. This amounts to the assumption that the causal mechanism, being roughly dominated by the interplay between mutation and selection, which is considered as responsible for the evolution on the species and sub-species level, also holds for the higher systematic categories (and morphological formations). However, several arguments have been stressed from independent sides against this view, especially from some people participating in the famous Serbelloni Symposia. A reflection, based upon probability calculations made by SALISBURY (Nature 224, 342, 1969), seemed to bring about the proof that even at the level of DNA the probability for the occurrence (or emergence) of self-reproducing autocatalytically active molecules is by far too low, i.e. by a tremendously high order of magnitude, for seriously being taken into account as an explanation for the origin of life and its evolution. These and similar results seem to obligatorily induce the inference that _form_ (or shape) has to be appointed as a prerequisite for possibly constraining probabilities and hence directing evolution. THOM's own metaphor of a moving wavefront reveals also a global view apparently surpassing local determinism as the ultimate reason for evolution.

COHEN (also a comment on the paper by ROSEN):
The general problem is similar to that discussed in my note. It is the evaluation of the real rate of evolution, especially that which brings about a fundamental change in the whole pattern of structure or behaviour of the organism. The perplexing question is what is it that makes one evolutionary line very successful in adapting and generating novelties in the long run, compared with other lines which die out or stagnate. In the terms of BREMERMANN, how does a species escape from an isolated local peak, or from being hung up on an extremely complex multi-dimensional 'ridge'. - In general, when a new ecological space is suddenly opened up, competition is greatly reduced for a while, and this will tend to connect up many previously isolated adaptive peaks, with the result that evolutionary frexibility is suddenly restored. In

such a situation, the species which had previously been closest to the now suddenly opened new pathway of evolution, will be the one which will be the first to occupy it, and to provide the source for all future evolution in the new space. It may be a matter of pure chance which one of several similar species is better pre-adapted to the occupation of a particular new ecological space. The chances are better however for relatively unspecialised species, as these have maintained a wider range of evolutionary options to cope with new situations. - Evolutionary frexibility is itself the result of the selections which have operated in the history of the species. Large and relatively frequent changes in the environment would have forced a continuous shifting in the genetic make-up of the species, and thus would favour those genes which require less stringent and specific interactions with other genes in order to function successfully. But this itself allows a faster response to further changes in the environment. The opposite is the case for a long history of a very stable environment, which would have favoured those genes which depend on specific interactions with many other genes. The result of this would be a population which can only change relatively slowly following changes in the environment. -

The most likely species to be the sources of further and novel evolution are thus the relatively unspecialised species living in environments which have recurring and relatively large long term changes. The most likely structures to form the basis of a novel development are those which have been released from strong and constraining interactions with other structures. For example, the evolution of the fore limbs to specialised wings and hands in birds and man respectively, following the assumption of bipedal locomotion.

GABEL:
BREMERMANN rightly contends that "species ... are not optimal in a global sense but only in a local sense" and "correspond to states that have been captured by attractors (ecological niches)". In agreement with these conclusions, CLOUD (1) has pointed out that the biological evolution of photosynthetic organisms seems to have been dependent on the prior existence of such ecological niches. The plethora of experiments which have been carried out for the purpose of studying abiotic, chemical evolution would also seem to indicate the necessity of attractors (microenvironments) either of prior existence or whose formation occurs concomitantly with the formation of new chemical species (2) (3). An operational definition of microenvironments can be given as a system in which a mass or energy parameter becomes limiting or dominating which then results in a predominant set of chemical reactions or physical states (including a repeating sequence of change of physical state) (3). Some geophysical examples of such microenvironments are areas of volcanic activity, saline ponds, brackish tidal areas, thermal springs, and mineral or hydrocarbon deposits. It should be borne in mind that DARWIN would not have conceived of the idea of biological evolution without having had the opportunity to observe microenvironments. - It is biologically observable that as each species or state becomes positioned by its attractor, other ecological niches or attractors are created or nullified by interactions between functioning attractors. This phenomenon is paralleled by the interaction of electronic energy states on a molecular level. Pluralistic phylogenesis and societies may in this sense be the expression of global optimization; and the sought after tool for the analysis of trajectories may be expressed in the definition of value (4).

References

1. CLOUD, P.E., Jr.: Science 160, 729, (1968).
2. PONNAMPERUMA, C. and GABEL, N.W.: Space Life Sciences, 1, 64 (1968).
3. GABEL, N.W. and PONNAMPERUMA, C.: (1971)"Primordial Organic Chemistry", In: PONNAMPERUMA, C.: (Ed.) Exobiology, Chapter 4, Amsterdam: North-Holland Publ.
4. GABEL, N.W.: this symposium

BREMERMANN:
A key problem for man and nature alike is <u>transcomputability</u>. The laws of nature generate an enormous space of possible systems trajectories, but neither nature with molecules and organisms, nor man with brains and computers, can explore more than an infinitesimal fraction of it. Only a minute number of all possible species can ever come into existence. In view of transcomputability the notion of <u>determinism</u> becomes problematic. The moment a system with a state space and with dynamic laws has been established all the possible trajectories are determined, but their knowledge may be inaccessible, to nature and to man alike.

LOCKER refers in his comments to a <u>local determinism.</u> If the term means that the values of certain functions need to be known only locally in order to compute the systems trajectories locally, then the possibility that the system trajectories are nevertheless determined <u>globally</u> would not be excluded. In fact many uniqueness theorems for the solutions of differential equations may be interpreted in this way. Also, I think that the question of local determinism is independent from the statement that species seem to satisfy only <u>local optimality.</u> Concerning the SALISBURY type of calculations to which LOCKER refers, I would like to call attention to Manfred EIGEN's paper (Naturwiss. 58, 465, 1971) who has argued against the assumption that go into these calculations and hence against the conclusions drawn from them. EIGEN has demonstrated the possibility of selfreproducing autocatalytic polynucleotides theoretically and he makes experimentally testable predictions for their chemical evolution.

COHEN's remarks concern the very complicated question of changing environment. This situation can be formulated abstractly (and our framework contains it) but it is even more transcomputational than the more restricted formulation of an environment that can be described by a type of <u>potential function</u> which is merely subject to perturbations. Nevertheless this question is very important and I think the principle of the selection of non-specialized species under changing environments is a very important one together with the principle of minimal interaction.

GABEL says in his remarks "An operational definition of microenvironments can be given as a system in which mass or energy parameters become limiting ...". I would like to point again to Manfred EIGEN's paper who has elaborated this point quantitatively for the "abiological" (chemical) evolution of polynucleotides and who has derived concrete results that can be tested experimentally. Indeed, it seems highly desirable that the theory of evolution should come up with explicit predictions that are verifiable experimentally.

The Limits on Optimization in Evolution
D. Cohen

The notion of optimality in evolution, following the Darwinian process of the "Survival of the fittest", is very useful when attempting to understand the way living organisms are built and behave. One has only to think of the many beautiful and precise adaptations to their environment which living organisms demonstrate. If we try to consider every species as being the product of a long process of evolution by natural selection, there are often, however, many cases where the performance of some particular structure or behaviour is obviously far from maximal.

In this short note I shall try to consider several limits or constraints on the process of optimisation during evolution.

1. Fitness is very often a very complex function in the space of all the structural and behavioural variables of an organism. Its function very likely has quite a large number of local maxima, separated by more or less wide regions of a lower fitness. Since evolution by natural selection is analogous to a search for peaks in a landscape in total darkness, knowing only the local gradient, it is not surprising that many of the peaks reached by an evolving population are relatively low. But as long as the landscape remains the same, there is absolutely no way for a population to get off a local low peak on which it had been trapped.

Selection does operate however to maintain the population at its local peak, even when the peak is changing its location, as long as the peak remains isolated. In this restricted local context, it is therefore quite legitimate to speak of true optimisation, and to apply predictions based on optimality considerations.

2. Different performance functions are very often correlated, for example by being dependent in a different way on the same body parameter. This correlation imposes a constraint on the optimisation of these functions, so that they cannot be achieved independently. The constraint often takes the form of a trade-off relation, so that any improvement in one performance function necessarily means a decreased performance in other functions.

Since the number and nature of these trade-off relations is usually only imperfectly known or understood, it is very difficult in most cases to calculate what is the optimal combination of such related functions. In fact, this difficulty is usually so great that a study of the optimal properties of an organism is only possible when these properties are independent from each other, or at most have well defined relations with other properties.

3. The frequency of mutations or recombinations which bring about an improvement in performance of any one function is usually quite low. The frequency of genetic changes which result in the simultaneous improvement of two interacting functions may be taken to be approximately the square of the value for the single function, and the same holds for higher numbers of interacting functions. It is clear that the frequency of a simultaneous improvement in several interacting functions becomes vanishingly small even for a moderate number of such functions.

A small change in the environment would be followed therefore by a corresponding change in the genetic make-up of the population, which will be fastest for the non-interacting functions, and progressively slower for functions with an increasing degree of interaction. Assuming that the probability of an improvement per generation for one non-interacting function is ca. 10^{-4}, it is easy to calculate that for 10 interacting functions it would be 10^{-40}, which means that even for a relatively large population of 10^{30}, there is only 10^{-10} chance of such a change appearing in any one generation in the whole population.

Unless the environment is inordinarily stable, the short term rate of change in the environment is faster than the rate of optimisation of complex interacting functions. This means that the present day populations of living organisms are mostly not at their short-term selection equilibrium, but are more likely on a rather flat fitness plateau, possibly quite a good distance from the nearest peak.

The usefulness of the optimality model is restricted therefore to non-interacting or weakly interacting functions. As the rate of change of the whole large complex of interacting functions is so slow, this complex can be taken as a constant basic framework within which the optimisation of the non-interacting functions takes place.

If the environment is mainly fluctuating around some long term mean, we would expect the fast changing non-interacting properties to follow fairly closely the short term rapid changes in the environment. The slow changing complex of interacting functions would follow the very long term changes in the long term mean of the environment.

4. <u>Population</u> vs. <u>Individual Optimisation</u>. Selection operates on the genetic contribution of an individual within a population, and on the fitness of the population relative to other populations of the same or other species. Maximal population fitness cannot be reached however, if the selection within the population favours those characters which are incompatible with this maximal fitness. Similarly, a gene favouring the individual carrying it, but destructive for the whole population, will be eliminated when the population carrying it is wiped out.

A fairly complex equilibrium may result when there are opposing selection pressures at the population and at the individual levels. Thus, optimality models based only on individual or on population optimisation criteria stand a good chance of being only partly correct.

Physical Problems of the Origin of Natural Controls
H. H. Pattee

Abstract
The physical embodiments of a control element or system are called non-integrable constraints. A constraint, in general, requires an alternative description to the microscopic description of the system. This alternative description selectively ignores detail which corresponds in the physical system to some form of dissipation process which gives rise to the new coordinations of the constraint and a simplified collective behavior we recognize as function. The origin of such control constraints must begin with low selectivity and imprecise function and gradually sharpen up to high specificity and narrow, precise function. The central physical problem is to understand the necessary conditions for primitive molecular control elements to evolve complexity of coordination and simplicity of function.

The problem of the <u>origin</u> of control is quite distinct from the problem of the <u>operation</u> of a well-defined control system. The basic physical distinction is that the operation of a control system assumes the existence of organized structures which are treated as boundary conditions or equations of constraint, whereas the origin problem must account for the establishment of these boundary conditions and constraints from the laws of motion starting from a relatively disorganized collection of matter. This distinction between the origin and operation of controls holds at all evolutionary levels. In fact it is the origin of these levels of hierarchical control which is the most difficult problem of evolutionary theory. However in this brief discussion we will emphasize only the simplest molecular level which was presumably the level at which life itself first originated.

The operation of controls in classical or statistical organizations or devices is now a well-developed mathematical theory, which does not directly concern us here (see, e.g., (1)). The basic idea of control theory is to produce a desirable or predetermined behavior in a physical system by imposing additional forces or <u>constraints</u>. The corresponding mathematical problem is either that of optimizing the control strategy under given constraints, or in determining the stability of the system with respect to the control variables or to outside perturbations. Most of these problems have their motivation in man's attempt to artificially control inanimate matter, other living systems, or to outwit his fellow man.

The problem I wish to discuss is the origin of controls that arise naturally, that is, without the intervention of man's intellect. These are the controls characteristic of living matter, and are embodied at the most fundamental level in copolymer molecules such as polypeptides and polynucleotides. Specifically, I would like to know to what extent the laws of physics can account for the spontaneous origin of molecular control systems.

What Is a Control Constraint?

The first problem in discussing the origin of natural control molecules is to distinguish them from all other molecules. We assume descriptive reducibility, i.e., that control constraints are microscopically describable by the laws of physics, and that at least in this respect they are not different from all other physical interactions or structures. In other words, we assume that a control system can be described in principle as completely as is possible for all matter using the same universal physical laws.

This brings me to the first essential point I wish to make: Since we assume that a control system is not distinguishable by its microscopic description we conclude that a control constraint requires an <u>alternative description</u> which for some purpose is more useful than the microscopic dynamical description. This is true for all constraints (9). The chemical bond is perhaps a significant example. In quantum mechanics a molecule can be treated as one stationary state of a collection of elementary particles called a system. The particular state this system happens to be in is incidental to the dynamics of these particles. In other words, in the quantum mechanical description no molecule is considered as a constraint on the particle dynamics. In fact, molecules can be called the results of the dynamics. On the other hand, there is the well-known alternative chemical language in which these stationary states are treated as essential constraints called chemical bonds.

Many other molecular aggregations are treated as constraints provided there is some degree of regularity that can be usefully described in a simpler alternative form. The most common regularity is a more or less persistent approximation to a geometrical surface such as a crystal face or an interphase boundary. In these cases one alternative description is simply the geometric equation for this boundary which ignores all degrees of freedom and hence all the dynamics of the elementary constituents forming the boundary.

It is clear however that our concept of control includes more than a fixed constraint or structure which permanently limits the motions of the system. A control must select between various alternative trajectories of a system. We may therefore propose the following definition: <u>A control device is a time-dependent constraint which alters the path of selected degrees of freedom of the system it controls in a variable but regular way</u>. In mechanics such path-dependent, non-integrable relations between variables are called non-holonomic constraints (e.g., (13)). The basic effect of such constraints is to make the number of dynamical degrees of freedom less than the number of coordinates necessary to specify the configuration of the system. It is because of these "extra" non-dynamical degrees of freedom that we may speak of controlling a system which otherwise would be completely deterministic. However, we must continue to assume that if we were to treat all degrees of freedom, including the constraining structure's degrees of freedom, with microscopic detail they too will follow the deterministic laws of motion and no clear distinction could be found between the control variables and the rest of the system being controlled. Therefore it can only be in terms of the <u>alternative description</u> (i.e., the equations of constraint) that the control aspects become a distinctly new property of a physical system.

From these assumption we are led to the second essential physical property of control constraints. The first assumption of descriptive reducibility of the dynamics of a control constraint includes the idea of the <u>completeness</u> of the microscopic description. In other words, we assume that the microscopic level is as complete a description of all degrees of freedom as is physically meaningful (i.e., no new degrees of freedom or hidden variables can be added). It therefore follows that any alternative description is either redundant, covering the same level of detail in an equivalent way, or else the alternative description is an abstraction covering <u>less detail</u> in certain degrees of freedom but realizing in return some <u>additional regularity in other aspects</u> of the system. Of course there are other forms of alternative descriptions in physics which ignore detail in return for additional regularity, the best known case being statistical mechanics. But this is not a control process since the loss of detail is a predetermined assumption on the part of an <u>outside</u> observer. This loss of detail and the resulting statistical regularity is not reflected in any detailed aspect of the system. For example, the pressure of a gas is a regular aspect of a collection which results from ignoring the detailed motions of the molecules; but without the addition of a rather complex control constraint this pressure exerts no effect on any of the detailed molecular motions.

The essential part of the control constraint is that part of the configuration which internally executes some form of averaging process over selected degrees of freedom while at the same time, through this dissipative process, establishes new correlations between other variables which then appear as the control variables. The laws of thermodynamics allow this trade at the well-known rate of at least kT of dissipation for each binary correlation. This same rate holds for any measurement or symbolic logical process which selects one of two equiprobable alternatives (e.g., (2) (7)). This relation is established for macroscopic devices, however a single quantum mechanical system acting as a control or measurement constraint is a much more obscure problem. In the first place, non-holonomic equations of constraints, when treated entirely formally in quantum mechanics, require operators which do not commute with the Hamiltonian and observables which have definite values as if a measurement had been made (5). This suggests that an even more basic difficulty is to understand what we could mean by an actual quantum mechanical control or measurement constraint. This in turn will require a more lucid quantum theory of measurement - a problem we cannot pursue here (e.g., (8) (10)).

However the basic point I wish to make with respect to the origin problem is that control constraints as well as measuring devices can be said to classify the microscopic degrees of freedom into sensitive and insensitive variables with respect to some collective behavior of the system. In most highly perfected control constraints this results in an extreme simplification over the detailed dynamics which we recognize as the function of the control element. The most basic example of an artificial control element which illustrates this functional simplification is the switch, while the enzyme molecule is perhaps the most elementary natural control element that performs a simple function even though its detailed dynamics are so complex that they are beyond the scope of present-day calculation.

Both the switch and the enzyme represent a highly effective and specific control element. That is, out of all the microscopic collisions with a switch or an enzyme, only very special ones are capable of triggering their catalytic function, and furthermore when this function is triggered only a very limited or simple result takes place. This is in contrast to ordinary dynamical systems where almost any collision results in a complex perturbation spreading through the entire system with no coherent result whatsoever.

The Origin Problem

Now we come to the origin question itself. The question needs to be broken into two parts: (1) How does any control constraint (e.g., a selective catalytic molecule) occur spontaneously, and (2) under what conditions do such control elements form coherent collections which persist and evolve? The second part of course is by far the more challenging since it is the evolutionary process that is the universal distinguishing property of life. We also know from direct experience over a wide range of both natural and artificial control systems that the elements of the system do not reveal, in themselves, the reason for their collective function, nor does a knowledge of the function determine any unique structure in the control elements.

This failure of the parts to reveal the collective function and the apparent arbitrariness of the embodiments of function have led many physicists as well as philosophers to a non-reductionist position, which amounts to the conclusion that control processes, while obeying physical laws, are not predictable from these laws (e.g., (11)). On the other side, the spontaneous appearance under primitive earth conditions of polypeptides and polynucleotides which look and act like they could become the control elements of cells have led many scientists so assume that integrated control systems are also spontaneous events which follow only from chemical laws (3) (6). I think it is not productive to follow these arguments here. What is missing from the origin problem are <u>testable hypotheses</u>, especially with regard to the second part of question, the origin of coherent behavior in collections that initially had no such coherence. This is a well-known central problem in evolution theory, the primary example being the origin of the genetic code.

The genetic code serves very well to illustrate the basic paradox of the origin of controls - that each element of the coding system is highly selective and simple in its function but loses all significance without the preexistence of all other coordinated elements. Thus no partial code seems to make sense, so that there does not appear to be any obvious gradual process leading to such a highly coordinated set of control molecules. On the other hand, there seems to be no reasonable likelihood of the sudden, spontaneous occurrence of such a complex, integrated collection of control elements.

Can our physical description of control constraints help clarify this problem or suggest a way out? We have emphasized the idea of a constraint as an <u>alternative description</u> of a physical system which can also be described in principle by microscopic laws of motion. This is therefore true of control systems in general. What is there, then, about the physical description of the origin of constraints that causes so much difficulty? In most physical theories we define the system we are talking about by fixing a number of degrees of freedom and then expressing the forces or interactions acting on these variables. It is easy to conceive of these forces being <u>gradually</u> removed so that we can see how the behavior of the fully interacting system grows continuously out of the non-interacting particles. We treat equations of constraint in quite another way - they are either present or absent - and there is no gradualness about it. This is because the equations are an alternative description. In other words, when we choose to use an alternative description there can be no gradualness about it. We cannot pass from one description of a system to another description in a continuous way. I believe this is the fundamental difficulty in describing the origin of new physical levels of control. All new hierarchical levels of organization require an alternative description (9) (12), and consequently there is no obvious way to describe a gradual transition between descriptions. Yet it is hard to believe that this is all nature's fault. The problem appears to reside largely in the nature of language itself which requires a discrete, fixed set of grammar rules which are not subject to continuous transformation. Again we may look at the physical interpretations to see if there is some way out.

We argued that an alternative description could only be consistent with the dynamical laws if it was redundant or resulted from a loss of detail which in turn required a dissipative process in the actual control element. We also expressed this process as a selection of sensitive and intensive variables with respect to the collective behavior of the system. Now these concepts - loss of detail, dissipation, selectivity, sensitivity - can be conceived in a continuous way, even though they must be discrete at the microscopic level. We also noted that highly evolved control elements such as switches and enzymes have a correspondingly high selectivity and highly limited function. A basic observed fact of effective control systems is that the functional simplicity can only be achieved through complex, coordinated constraints. It is precisely this complexity that makes the alternative description necessary, while it is the functional simplicity which makes the alternative description sufficient.

The origin of control constraints is therefore a different type of problem than we normally encounter in physical explanations where the complex system is described in terms of simple elements. The origin of control requires a description of how simplicity grows spontaneously from complexity. To be more precise we may think of complexity in terms of the number of degrees of freedom which can be followed in detail in a dynamical description, while simplicity is related to the number of variables left in the alternative description using the equations of constraint. But using the previous ideas of the physical nature of control constraints we see that whereas the language of the alternative description of the final function arises discontinuously, the growth of constraints must be gradual. It is therefore more useful to think of the development of constraining forces or of an <u>optimization of the degree of constraint</u> (PATTEE (9)).

In terms of the physical properties of constraints which we discussed, this means that there is a corresponding gradual optimization of the amount of lost detail or dissipation, selectivity, sensitivity, and range of output function. For example, according to this idea we would expect the proto-enzymes to have been very broadly specific with respect to the types of molecules which could act as substrate, and that the catalytic effect would be very weak with many subsidiary effects. The hypothesis also implies that this imprecise proto-enzyme function can be executed by a correspondingly broad class of copolymer sequences. We could imagine this primitive control situation as a blurred picture in which catalysts and substrates cannot be clearly resolved and therefore cannot be simply distinguished. At this level no useful alternative description of the complex dynamics would in fact be possible.

Almost all of the examples I know support this general process of the evolution of control as a gradual sharpening up of specificity and simplification of function, but unfortunately they are drawn from highly evolved biological systems in the first place so that they give little conceptual help with the origin of life problem at the molecular level. The origin of artifacts follows this course, originating with low specificity and broad function and gradually optimizing the degree of constraint, narrowing both the specificity of function as well as the physical structure which executes the function. The hammer is a simple example. We imagine its origin with such a loose classification of both structure and function that almost any object that could be picked up and used to strike any other object serves as a prototype. Then gradually we see both structure and function sharpened until we have dozens of highly differentiated and highly effective types of hammers. We see this course of evolution in so many biological systems, in differentiation, in speciation, in the development of symbolic systems and languages, that we usually take it for granted.

Yet this picture presents us with what I see as the fundamental origin problem. What are the physical principles or conditions which must be satisfied for the most primitive molecular control system to evolve? Why should a chaotic collection of weak catalysts with marginal specificity gradually sharpen their specificity and catalytic power and condense out coordinated control systems? Is there a threshold of specificity and reliability in the control elements necessary for the evolution of a control system? This is an area where I believe theory and computer simulation must play an essential role in understanding the origin of control systems, since it is the nature of highly evolved control systems to obscure their own origins, and primitive natural systems no longer exist. In this brief paper we have emphasized the physical requirements for control elements, but of equal importance is the structure of the environment from which these elements arose (3). However, this is a different aspect of the problem and must be so recognized, for the environment is conditional while physical laws are universal.

References

1. BELLMAN, R., KALABA, R.: Selected papers on mathematical trends in control theory. New York: Dover Publ. Inc. 1964.
2. BRILLOUIN, L.: Science and information theory, 2nd ed. New York: Acad. Pr. 1962.
3. CALVIN, M.: Chemical evolution, Oxford: Univ. Pr. 1969.
4. CONRAD, M., PATTEE, H.: J. Theor. Biol. <u>28</u>, 393 (1970).

5. EDEN, R.J.: Proc.Roy. Soc. <u>205 A</u>, 583 (1951).
6. KENYON, D.E., STEINMAN, G.: Biochemical predestination, New York: McGraw Hill 1969.
7. KEYES, R.W.: Science <u>168</u>, 796 (1970).
8. PATTEE, H.: J.Theor.Biol. <u>17</u>, 410 (1967).
9. PATTEE, H.: In WADDINGTON, C.H. (Ed.): Towards a theoretical biology, vol. 3, p. 107, Chicago: Aldine Publ.Co. 1971 a.
10. PATTEE, H.: In: BASTIN, T. (Ed.): Quantum theory and beyond, p. 307, Cambridge: Univ. Pr. 1971 b.
11. POLANYI, M.: Science <u>160</u>, 1308 (1968).
12. ROSEN, R.: In: WHYTE, L.L., WILSON, A., WILSON, D. (Eds.): Hierarchical structures, p. 179, New York: Elsevier 1969.
13. WHITTAKER, E.T.: A treatise on the analytical dynamics of particles and rigid bodies, 4th ed., p.214, New York: Dover Publ. Inc. 1937.

DISCUSSION

GABEL: Perhaps the problem of the origin of control, which PATTEE poses, is not quite so formidable as it would appear. Evolution's element of control which PATTEE is seeking is stated in the operational definition of excitability. "Excitability is the property of an organized group of particles to transfer environmental information along its parts resulting in the maintenance of its structural integrity (1)". The geophysical origin and the consequent corollaries of this definition will be found in the referenced paper from which the quotation was taken and in my paper (this symposium). The laws of physics are not violated; geophysical reality is not ignored; and the definition is, furthermore, operational for any system (evolving or not). The suggestion that "computer simulation must play an essential role in understanding the origin of control systems" might be valid if the computers were programmed to be responsible for the primacy of their own survival. I am grateful to PATTEE for underlining the importance of the origin of control.
GABEL, N.W.: Life Sciences, 4, 2085 (1965).

ROSEN: The problem of alternate descriptions raised by the author in connection with the origin of life problem is a profound one in many other areas of biological investigation (or indeed whereever one finds a complex system or organization). It appears even in engineering, in the distinction made between a "state variable" description of a system, and an "input-output" or "black-box" description of the same system. The input-output descriptions are most useful for constructing block-diagrams of the overall functioning of a system in terms of distinguished subsystems with definite functions, interconnected by oriented arrows representing flows of material and/or information. In these flows we clearly see the feedbacks and feed-forwards which are responsible for the way in which the entire system behaves. On the other hand, if we describe the same system by lumping together all of its state variables on a equal footing (as we tend to do in biochemical or population models) we completely lose the functional aspects and see merely a big system with some overall stability properties. A major problem, therefore, is to be able to dissect such large dynamical systems into diagrams of blocks or subsystems which are "controlling" each other. This is a problem which so far has hardly been touched in engineering or mathematics, but which has many ramifications in biology and elsewhere; I have some preliminary results for specific systems, which essentially boil down to the statement that such decompositions are highly non-unique. This by itself, however, is reminiscent of the existence of many low-order specificities simultaneously manifested by the same system, which can then be magnified selectively by local environmental conditions of which the author speaks.

WALTER: Why would each element of a code lose all significance without the pre-existence of all other elements? Couldn't the current status of a code be that the code is evolving toward a form that could contain additional information? Indeed, a partial code makes a great deal of sense! And surely the code of "tomorrow" need not be insignificant simply because it does not exist "today".

PATTEE: The present genetic code appears to be universal and shows no significant evolutionary change over a very wide range of species. If any one coding enzyme or transfer RNA were missing, then the corresponding amino acid would be missing from all other coding enzymes as well as all other replicating and transcribing enzymes. That would undoubtedly be lethal. – More abstractly, a "partial code" only makes sense to me if it is a sub-code, that is a complete code found within a larger code, like a sub-group is formed within a group. This closure property is what WADDINGTON might call an archetypal constraint in evolution – a constraint that does not depend only on fitness or adaptedness for survival, but rather on an intrinsic coherence.

BREMERMANN: The notion of _alternative description_ seems to me equivalent to the notion of _model_. A hierarchy of alternative description thus becomes a hierarchy of models. The main reason for working with simplified models rather than original (quantum mechanical) descriptions is the _mathematical and computational intractability_ of the latter. – Manfred EIGEN's new theory fo chemical evolution shows how nucleotide sequences could evolve without many of the control mechanisms of a modern cell.

PATTEE: The connotations of "description" and "model" are not equivalent in my mind. However, in some particular cases of hierarchical organization, I am sure that the concept of model is more explicit. Other concepts such as likeness, representation, theory, analog, metaphor, etc. might also be more explicit descriptions in other cases. – EIGEN's computer simulation is the type of work which I believe is a promising technique for testing theories of origin and evolution. I would also like to see realistically complex chemical experiments, which are not aimed only at synthesis of biochemicals, but designed for the study of primitive earth geochemistry and ecology (e.g., PATTEE, H.H., in: BUVET, R. & PONNAMPERUMA, C., Eds.,: Molecular evolution I: Chemical evolution and the origin of life, p. 42, Amsterdam-London, North-Holland Publ.Co., 1971).

FONG: I think the basic questions here (1) how does any control constraint occur spontaneously and (2) under what conditions do such control elements form coherent collections which persist and evolve, can be looked upon from the standpoint expounded in my paper. The questions arise when we consider control and coherence as a priori concepts but they disappear when we realize that they are a posteriori. My basic point is that life is an a posteriori manifestation of a process characterized by exponential multiplication. Any evolutionary changes in the microscopic description that occur according to the dictation of the fundamental laws of physics are purposeless and incoherent. Yet, those changes that make the exponential multiplication more efficient persist and overwhelm the others merely by virtue of population statistics (not by any a priori laws of nature). The process also sharpens the specificity as evolution proceeds. It is with respect to the more efficient exponential multiplication that the a posteriori concepts of control and coherence emerge. My view is not different from that suggested by PATTEE's discussion on the evolution of the hammer which is the essence of the paper; nevertheless it disposes of the conceptual problems elaborated in the rest of the paper.

PATTEE: The question I am asking does not disappear simply by considering life as an "a posteriori manifestation". In fact, the very problem is the origin of a posteriori modes of control, i.e., from observed effects to rules or causes. How do the events of nature come to be "observed" or recorded in the first place? It is only these hereditary records and the coherent constraints which read them out that make biological evolution possible. "Exponential multiplication" or differential survival are not sufficient for natural selection – there must be a genetic record as well as the mechanism for construction of the phenotype.

LOCKER: In my introductory communication to this Symposium I referred to your important quotation of non-holonomic constraints which convinces everybody that constraints are indispensable tools for constructing hierarchies. In this respect I would like to know your opinion about the mode of action by means of which constraints in general are able to exert their influences. In my mind any assumption of the existence of constraints by necessity requires the additional postulation of a kind of <u>meta-organization</u> behind (or beyond) that structural and functional organization that is primarily the subject of consideration. This meta-organization can, of course, be regarded as static or dynamic. An answer to this question, as you correctly state, seems to be in close dependence on the capabilities of language (or, what I would prefer to say, of intellect) to alternatively describe a situation. We must presume that the possibility of an alternative description does not only indicate the occurrence of hierarchies - that is absolutely correct - but also the inadequateness of the description as such. Wherever during the process of cogitating about a problem a transition (or transgression) of semantic steps is performed, then the inevitable task to appropriately approach the problem calls for an alternative description. However, in the long run, this description need not remain a purely alternative one, occurring at the same semantic level which the first description dwells on, but rather an additional (complementary) one as afforded by the view from another angle (or even another level). To continue the enquiry into this intricate matter: What enables one to alter the position or point of view? Granting that constraints are structurally determined one has to resort to those relations along which (or according to which) a shift from one point of view to another becomes possible. Hence, without the insertion of a self-like entity just at this very point it would not make sense to speak of an alternative (or <u>complementary</u>) description. The underlying relations offer a prerequisite for bringing about the transition but do not yet realize it. - There is a particular statement in your paper in which you deny the possibility of a gradual (continuous) transition between several modes of description on the one hand, such that any alternative description (i.e. the constraint) arises discontinuously, whereas on the other hand a continuous development of those constraint can happen until a certain optimization results. If this statement should not contain an intrinsic contradiction it must be understood in this way - that in the first instance a subject-like self has been tacitly introduced whilst in the latter instance an objective occurrence (and development) of constraints is described. - There is still another remark which fascinates me, namely the idea propounded by you that through the dissipation process, i.e. the loss of detail, new correlations between variables are established. This idea appears to be based on the presupposition of only a fixed number of correlations tolerable for any system. In case that some of them are dissolved (i.e. dissipated) others assume so-to-speak vicariously the role of the control entity within the system. Thus, the system is ascribed a certain constancy property which again constitutes a meta-organization.

PATTEE: You are posing the most fundamental questions and therefore the most difficult to answer. I would tentatively accept a coherent set of constraints as a language only if it possesses the self-referent or metalanguage capability - even though I cannot precisely define what I mean. Of course if I could define a metalanguage, then I presume I could define a language. Now when I deny the possibility of a gradual transition from one level of description to another I mean it in the same sense that I cannot pass from the object language to the metalanguage by a continuous transition. I am either talking about events or I am talking about the symbols which stand for the events - there is nothing in between that has much meaning that I can imagine. One could also use the analogy of speaking in two different languages, say French or German. At any moment I am either speaking French or German. Even though I may slip from one to the other for a word or phrase, there is no "intermediate" language. - On the other hand, I picture most elementary physical systems (at least, classical systems) as changing gradually or continuously. I can move interacting systems further and further apart and thereby gradually decrease the forces between them, or I can perturb them by an arbitrarily small force. When I introduce what I call a constraint, however, it has the effect of changing languages not simply changing or perturbing the forces.

I regard this "contradiction" as no worse a problem than the wave-particle duality in quantum theory, although that is bad enough. In fact, I would suggest that it will turn out to be more helpful to think of the wave-particle duality as well as the measurement problem of quantum theory as stemming from the matter-symbol relation rather than from purely physical principles. Whether or not matter is essentially continuous or discrete, I cannot imagine symbols or classifications as other than "lumping" processes which are functionally discrete.

The Significance of Cooperative Interactions in Biochemical Control Systems[+]

Ch. Walter

Abstract [++]

In classical enzyme kinetics a hyperbolic relationship between the initial rate and the initial substrate concentration does not depend upon whether or not the rate of turnover of the enzyme-substrate compound is slow compared to the reverse of the enzyme and substrate binding reaction. However, in enzyme mechanisms involving ternary or higher order compounds it is sometimes possible to obtain non-hyperbolic relationships only if the type of quasi-equilibrium described above is not maintained. We examined the effect on the sigmoidal relationship expected from the classical "allosteric" model when the context is an enzyme that turns over rapidly (rather than a protein that simply binds a ligand). The results indicate that the sigmoidal shape of the relationship between the initial rate and the initial substrate concentration can be destroyed under certain general circumstances: 1) If the rate of turnover of the enzyme-substrate compounds formed from the favored form of the free enzyme is less than the rate of turnover of the other enzyme compounds, a hyperbolic relationship (with slight or no substrate inhibition) rather than a sigmoidal relationship can result whether or not quasi-equilibrium is maintained. 2) If the rate of the reverse of the enzyme and the substrate binding reactions for the favored form of the enzyme is considerably faster than the rate of the reverse binding reaction for the other form of the enzyme, a hyperbolic rather than a sigmoidal relationship can result provided quasi-equilibrium is not maintained.

LOCKER: You very nicely describe the mechanisms by means of which a prevalence of the hyperbolic over the sigmoidal relation between rate and initial substrate concentration is achieved. However, in the context of our Symposium it would be extremely interesting to ponder on the relevance of such mechanisms for the evolution of copolymers which have arisen and could have found a selective advantage by using a special enzymatic mechanism that turns out to be more effective than another.

WALTER:
Final remark not received.

[+] This work was supported in part by National Science Foundation Grant GB 20612.
[++] Full paper will presumably appear in BG&T.

Organization of Glycolysis

B. Hess

Abstract+

A self-coupled and cross-coupled network of chemical interactions ensures the detailed control of the sequential operation of the glycolytic enzyme sequence. The control range allows a wide variation of activation, inhibition and oscillation dynamics of the chemical flux through the system. Within the living cell the glycolytic enzyme system is closely packed, thus minimizing the transit time of glycolytic intermediates. The kinetic parameters of the individual enzymes can be ordered in allowing transient times in the range of 10^{-3} to 10^{-1} sec of individual enzymic steps. The enzymes are only saturated to a small degree of 0,01 - 0,2 allowing a powerful conformation control of the regulatory enzymes by controlling ligands as well as a first order function of the non-regulatory enzymes. This property is reflected by the subunit structure of the enzymes which not only ensures cooperation of control functions but also seems to provide a spatial advantage. - It is interesting to speculate on the evolution of such an organization. In the evolution of higher organisms from lower ones the total cellular space increases. This certainly implies diffusion problems which might be managed through the organization of oligomeric structures of enzymes.

+ Full paper will presumably be published in BG&Th.

Cell Models and the Homeostasis Problem
Z. Simon

Abstract

In mammalian cells initiation of DNA replication requires RNA synthesis in the early G_1 - phase and synthesis of an inducer during the S - phase. The mitotic operon may be switched on and off by an operonic trigger which in turn is based upon the mitotic operon and the histospecific operon. The assumption of a block of the two operon activities during the S - phase guarantees the oscillatory action of the trigger. The model contains the feedback with the total cell number via the repressor for the histospecific operon. Taking into account a competition for micromolecular precursors between the processes implied in protein synthesis and enzymatic reactions, a single micromolecular precursor for protein synthesis suffices which is synthesized under the control of an enzyme (as far as steady state can be assumed) and for whose decay a lytic enzyme is required in addition. The amount of gene activation is determined by the switching states of certain triggers. For all-or-none triggers one may adopt binary (logical) functions (2). The gene activities which enable the different kinds of work done by stem cells (low proliferation), immature cells (high proliferation), mature cells (high histospecificity) and dying cells, can be obtained by a system of triggers. In the model of cell cycle the cell volume must be included, too. The initiation of DNA replication can be considered as due to accumulation of a threshold membrane substance. The feedback between growth rate and the numbers of j-type cells (i.e. stem-, immature-, mature or dying cells) can be considered as following (3) with intercellular diffusion of the chalone. A pair of equations is obtained which describes: 1. the intra-extra-cellular exchange of chalones for cells of this type and 2. the balance of the extracellular chalone being subject to a decay of certain rate. The model neglects many aspects of cell regulation. However, since DNA transcription is the first amplifying cascade in the sequence of DNA $\longrightarrow$ RNA $\longrightarrow$ enzyme $\longrightarrow$ enzymatic reaction the consideration of gene activation control can be regarded only as a first approximation which must be elaborated further.

References

1. SIMON, Z., FARCAS, D., CRISTEA, A.: In: LOCKER, A. (Ed.): "Quantitative Biology of Metabolism", p. 71, Berlin-Heidelberg-New York: Springer 1968.
2. SUGITA, M.: Rep. Progr. Polymer Phs. Japan 10, 541 (1967).
3. TSANEV, R., SENDOV, B.: J. Theor. Biol. 23, 124 (1969).

DISCUSSION

ROSEN: The control of the size of an aggregate of cells through a control on the proliferation rate of the individual cells in the population raises numerous difficult questions. A model similar in spirit to the one considered here, which makes an interesting comparison, is that of YCAS, SUGITA & BENSAM (J. Theor. Biol. 9, 444, 1965,) cf. also ROSEN (J. Theor. Biol. 15, 282, 1967) for a more general theoretical setting of this model. It also occurs to me that such models can be regarded as special cases of the kinds of "positional information" mechanisms which have been proposed by WOLPERT (J. Theor. Biol. 29, 147, 1970) and others to account for aspects of differentiation and pattern generation during development. Indeed, I think the author's model can be regarded as a general positional information scheme, simply by regarding a more general kind of next-state decision of a cell than simply whether to divide or not.

SIMON:
Final remark not received.

Contribution to a Mathematical Theory of Synergic Systems[+]

N. A. Coulter, Jr.

Abstract
The theory of non-interacting controls, designed to eliminate interactions among control channels in a complex system, is well known. From the standpoint of the engineer, non-interacting controls are desirable. Interactions among the components of living systems, however, not only exist, but frequently possess the property of synergy – i.e., they are mutually reinforcing, in some sense. In this paper an outline of a mathematical theory of synergic systems is presented. Such a theory is basic to an understanding of teleogenetic systems – systems endowed (genetically or otherwise) with the capacity to generate their own goals.

The work presented here emerged in the course of research on teleogenetic control systems (1) – systems which are not only goal-seeking or teleonomic (PITTENDRIGH (2)), but which also have the capacity to generate their own goals. It was found necessary to consider interactions among some of the components of such a system.

The theory of non-interacting controls is well-known (3). The problem confronting teleogenetic system theory, however, is how to determine the conditions under which interactions are <u>synergic</u> – i.e., mutually promoting, in some sense, the interacting components.

To illustrate what is meant here, let us consider a system of two components, as shown in Fig. 1.

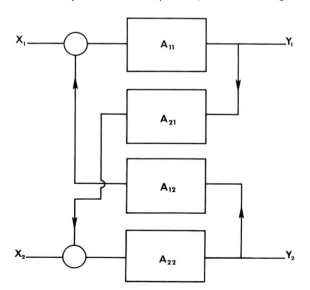

Fig. 1.
System of two interacting components

[+] Supported by N.A.S.A. Grant NGR 34-003-040.

Assuming each component is linear, the behavior of the system is governed by:

(1)
$$Y_1 = A_{11}(X_1 + A_{12} Y_2)$$
$$Y_2 = A_{22}(X_2 + A_{21} Y_1)$$

Solving for Y_1 and Y_2 respectively, we get

(2)
$$Y_1 = \frac{A_{11}}{1 - A_{11} A_{12} A_{21} A_{22}} X_1 + \frac{A_{11} A_{12} A_{22}}{1 - A_{11} A_{12} A_{21} A_{22}} X_2$$

$$Y_2 = \frac{A_{11} A_{21} A_{22}}{1 - A_{11} A_{12} A_{21} A_{22}} X_1 + \frac{A_{22}}{1 - A_{11} A_{12} A_{21} A_{22}} X_2$$

In the characteristic engineering system, the "value" or "worth" is external to the system - determined either by engineer or user, or by some social process. In a teleogenetic system, however, the system itself assigns worth-values according to some algorithm or heuristic policy. To accomplish this, a teleogenetic controller has to be superimposed upon the basic system. One such control loop is diagrammed in Fig. 2. The teleogenetic controller consists of two components, an Evaluator and a Director.

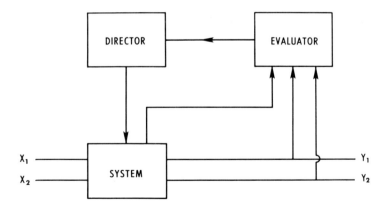

Fig. 2. Teleogenetic controller of two component system

The Evaluator assigns worth-values to Y_1 and Y_2, which it transmits to the Director. The Director, a decision-making component of the system, modifies the transfer characteristics of the system.

To continue in more detail, let us assume that the Evaluator performs the following operations:

1. It computes worth-values according to the functionals

(3)
$$W_1 = \frac{1}{T} \int_o^T K_1 Y_1 dt$$

$$W_2 = \frac{1}{T} \int_o^T K_2 Y_2 dt$$

where T = epoch of evaluation
K_1, K_2 = Kernels, functions of t

2. It computes cost functionals on the interaction transfer functions A_{12} and A_{21}

(4)
$$C_{12} = L_1 \left[A_{12}(s) \right]$$

$$C_{21} = L_2 \left[A_{21}(s) \right]$$

where L_1 and L_2 are functionals

3. It determines, after discrete epochs, a global utility representing numerically the relative worth and cost values:

(5)
$$U_T = \frac{W_1(T)}{C_{12}(T) + 1} + \frac{W_2(T)}{C_{21}(T) + 1}$$

where U (T) = global utility for the epoch T.

Note that for zero cost functionals, associated generally with zero interaction transfer functions, the global utility is simply the sum of the worth functionals.

It is assumed here that the direct transfer functions A_{11} and A_{22} are fixed by the physical characteristics of the system. The interaction transfer functions, however, are subject to Director control.

The Director is a computer, which has the capability for switching the interaction transfer functions A_{12} and A_{21} for the epoch T + 1, according to a decision based on the global utility associated with the epoch T. (We assume a "dead time" τ between epochs to permit Director computations to be made and transients associated with shifts of A_{12} and A_{21} to subside. Thus epoch T + 1 does not start immediately after the end of epoch T.) The question arises: what decision criterion should be used by the Director in its selection of interaction transfer functions?

Intuitively, it seems appropriate for this criterion to be such that the global utility function would be increased. A variety of such criteria are possible. We will confine ourselves here to a very simple criterion, based on the assumption that the system inputs X_1 and X_2 are stationary.

The Director compares the null-cost global utility with global utilities associated with a particular set of interaction transfer functions. The null-cost global utility U_o is

(6) $\quad U_o = W_1 + W_2$

The Director then chooses a set of interaction transfer functions and computes the global utility U for those transfer functions. The difference,

(7) $\quad \Delta U = U - U_o$

is then determined. When ΔU is positive (or exceeds some pre-set threshold), the system is synergic for the given set of conditions.

In the simplest case, the Director is programmed to compute global utilities for different interaction transfer functions until a synergic set is found (ΔU is positive). It then switches the actual interaction transfer functions to this synergic set.

DISCUSSION

The basic purpose of this paper is to draw attention to the possibility of synergic systems and to demonstrate the feasibility of the development of a mathematical theory of such systems. Clearly, only a beginning has been made here. A number of questions, problems, and potentialities suggest themselves for consideration.

First, this particular synergic system is obviously one of a large class. In addition to the extension to multi-component and nonlinear systems, there is nothing unique or necessary in the choice of worth-functionals, cost-functionals, or global utilities. It is reasonable, but not necessary, to associate the worth-functionals with the system outputs; one might with equal justification associate them with one or more state variables. It is reasonable, but not necessary, to associate the cost functionals with the interaction transfer functions, since they are under Director control.

Second, the question might be raised, why not use an optimum ΔU as a criterion for selection of interactions transfer functions?

The answer to this is that of course an optimum ΔU would be desirable. The appropriate application of optimization theory is highly relevant to the development of a mathematical theory of synergic systems. At this stage, however, certain points merit consideration:

1. An optimum synergic system is a particular case of a synergic system. A general criterion for distinguishing synergic systems from the set of all possible systems is desirable. There is no assurance that an optimum synergic system exists, even for the particular case described here.

For this particular case, another important consideration is computation time. Intuitively, it would appear that the computation required to find a synergic system would be less than that required to find the optimum system. Since prolonged dead times are undesirable, this consideration is of practical importance.

2. A major problem is: how should the director generate new interaction transfer functions for computing ΔU? There does not appear to be a simple best solution to this problem. Among the policies possible are:

a) A random selection.
b) A systematic selection in some definite order.
c) A heuristic selection, based for example on choice of interaction transfer functions which were synergic in the past.

3. It has been tacitly assumed here that a synergic system exists, for any particular set of inputs, worth-functionals, and cost-functionals. A general proof for this has not yet been found. However, it is easy to show that synergic systems can exist, and if none exist for certain non-synergic combinations, the obvious decision for the Director is to choose zero interaction transfer functions (or functions for which cost-functionals are zero).

Finally, the potentialities of a mathematical theory of synergic systems merit consideration. While there is as yet no demonstration that any living system is synergic (or at least sometimes synergic), the question has never really been asked in any quantitative sense. Qualitatively, living systems do appear to seek to enhance "pleasure" and to avoid "pain", and the ideas presented here do provide a quantitative basis (not the only one possible) for characterizing systems which exhibit these "qualities".

Quite apart from its possible application to biological systems, a mathematical theory of synergic systems may possibly be of value in dealing with ecological and social problems. In view of the urgency of these problems, it seems desirable to call attention to this potentiality.

References

1. COULTER, N.A., Jr.: "Toward a theory of teleogenetic control systems." General Systems Yearbook, 13, 85 (1968).
2. PITTENDRIGH, C.S.: In: Behavior and Evolution, ROE, A., SIMPSON, G.G. (Ed.), p. 391, New Haven: Yale University Press, 1958.
3. BOKSENBOM, A.S., HOOD, R.: "General algebraic method applied to control analysis of complex engine types." NACA Rep. 980, April, 1949.

DISCUSSION

ROSEN:
Theories of control, and optimal control, always assume that we have a specific system which we wish to control, in accord with goals and constraints imposed from outside the system. To accomplish this a variety of accessory systems are imposed, and allowed to interact with the system to be controlled, in such a way that the controlled system exhibits the desired behavior. On the other hand, when we look at the controlled system, together with its auxiliary controllers and evaluators "from outside", as it were, it tends to look to us like one big system, with the control aspects disappearing into the overall dynamical properties of the system as a whole. Thus, the applicability of these control theories to large systems of unknown properties is very much less obvious than it would initially appear. - Conversely, if we do have a big system, it is possible to split it apart into subsystems (usually in many ways) which do interact with each other in the fashion of a controlled subsystem with its associated controllers. Such decompositions are highly non-unique, of course, but in the analysis of complex biological systems it might be possible to evaluate each of them in terms of the kinds of efficiency criteria suggested from engineering practice, and which the author has outlined in his paper. From this we might be able to identify some "best" or "optimal" decomposition, which would allow us to say that it is this optimal decomposition which embodies the actual function or activity of the system, in terms of its information flow.

COULTER:
I quite agree. This raises the interesting question of how to identity this optimal decomposition. MESAROVIC' Principle of Uncertainty in structure of a multivariable system is applicable here. It also seems to be that there may be no single "best" decomposition, that there may be multiple optima. If this is so, there may be alternative modes of comprehension, equally valid for explaining system behavior and for predicting its future course.

LOCKER:
Does it suffice for a true teleogenesis that only worth-values are computed by the evaluator? It seems to me that the operations you describe can only be performed under the assumption that the goal is in principle already tied with the system's organization and that for a special task certain adjusting operations would select that behavior that is appropriate in enabling the system to act according to the goal. Thus, the goal lies within the scope of the system and is not really created (or generated). Another question: What consequences emerging out of your considerations do you see for a theory of biogenesis?

COULTER:
The goal is tied to the system's organization in the sense that (1) the structure of the system limits it to an ensemble of goals, an ensemble which, however, may be quite large (2) the stored "experience" of the system further limits it to the selection of a subset of the ensemble, a subset which may also be quite large. The situation is analogous, I think, to that of programs available to a computer. The instruction set wired in to the computer limits it to those programs that may be generated using those instructions, and storage limitations and time and energy requirements restrict the choice of programs still further. Nevertheless the number of possible programs is quite large. I did not mean to imply that goals were created de novo, but were generated from an ensemble of possible goals. One possible consequence of teleogenetic system theory for the problem of biogenesis is that it suggests a possible criterion for the emergence of living systems, namely that the organization of the system is sufficient to permit teleogenesis to occur. Even granting the validity of such a criterion, however, teleogenetic system theory is not yet sufficiently developed to permit the criterion to be specified.

The Role of Precursors in Stimulating Oscillations in Autocatalytic Diffusion Coupled Systems

B. Gross and Y. G. Kim

Abstract

This paper examines the relationship between the structure of a biological system and its oscillatory behavior through analog simulation of two "reactor systems" in which a chain of autocatalytic reactions could take place. The role of intracellular compartmentalization in cell function and the interaction of the cell with external stimuli are discussed herin.

INTRODUCTION

The appearance of oscillations in an open system in which chemical reaction and mass transport interact has particular significance since biological cells may be regarded as such systems. The biological cell is basically a reaction vessel containing all necessary catalysts, into which metabolites are free to flow and therein chemically react.

The properties of the open system are such that a dynamic steady state is possible, with the energy and/or entropy of the cell increasing at the expense of the environment. Cell growth as well as drug adaptation are explainable in terms of an open system.

The relationship between the structure of a biological system and its oscillatory behavior is becoming increasingly important with the discovery of periodicities in enzyme reactions, cellular studies and complex reactions of mitochondria.

Several instances of oscillatory kinetics have been observed (3) (4); for some of the oscillatory mechanisms the assumption of autocatalytic reactions has been made (2) (5) (7). An autocatalytic chain of reactions has several properties, in addition to its tendency of periodicity, that make it an attractive tool for simulating biological systems. For one, a chainlike series of reactions is a characteristic of many metabolic pathways. For another, many enzymatic reactions are of the autocatalytic form.

Varied levels of structure or localization are common in biological systems. The chain of reactions was, therefore, studied in both the structured (compartmentalization) and unstructured (homogeneous) configuration. The homogeneous model, in a more restrictive form, has been discussed (9). The differential equations governing the compartmental model have been presented without attempting to simulate them (1).

The structured state was simulated by restricting each reaction in the chain to a specific site, or compartment, within the cell. Each site contains a specific catalyst or enzyme and thus only one reaction can take place in each compartment. Species will diffuse to the site, react, and product will diffuse to the next site and in turn react. Each specie is also capable of leaving the cell via diffusion. The unstructured state was simulated by allowing each reaction step to occur uniformly within the cell boundary. Of key interest in this study was the systems response to outside stimuli, the role of oscillations in the response, the role of intracellular compartmentalization in cell function, the interaction with external stimuli or precursors, and the way in which various levels of intensive variables affect system behavior. Some of these effects have been discussed previously (6).

DESCRIPTION OF THE MODELS

1. Homogeneous (Unstructured) Model

The system of n species undergoes reaction in a cell which is approximated by a well-stirred reactor (Fig. 1). All reactions are autocatalytic expect the last which is first order. Each specie

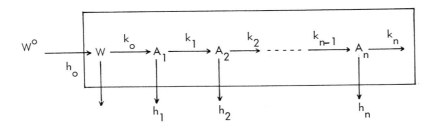

Fig. 1. Schematic Diagram of Compartmental Model

may diffuse out of the cell, which has permeability h_i toward specie A_i. Incorporating the rate equations into a mass balance the following set of differential equations was obtained (9).

$$(1) \quad \begin{aligned} \frac{dA_1}{dt} &= K_o W A_1 - K_1 A_1 A_2 - \sigma_1 A_1 \\ &\vdots \\ \frac{dA_i}{dt} &= K_{i-1} A_{i-1} A_i - K_i A_i A_{i+1} - \sigma_i A_i \\ &\vdots \\ \frac{dA_n}{dt} &= K_{n-1} A_{n-1} A_n - K_n A_n - \sigma_n A_n \end{aligned}$$

$\sigma_i = h_i \, S/V$, in which S/V is the surface area to volume ratio of the cell. The symbols W, A_1, A_2, ... are used to represent concentrations of the species W, A_1, A_2, ... as well.

Two entry conditions determine the behavior of precursor, W.

(a) Concentration of W is constant in the cell

$$(2) \quad \frac{dW}{dt} = 0$$

(b) Precursor enters by diffusion from the medium where its concentration is maintained at W_o

(3) $$\frac{dW}{dt} = \sigma_o (W_o - W) - K_o W A_i$$

2. Structured (Compartmental) Model

As an alternate to the unstructured model, it is assumed that the cell contains sites of enzyme localization, so that only one reaction will take place in a particular section of the cell and product must diffuse to another site before it can react. Again each specie is formed autocatalytically and may also diffuse out of the cell. Schematic representation is given in Fig. 2.

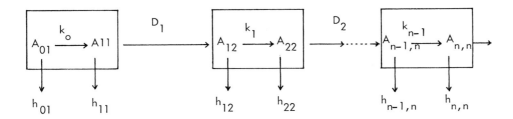

Fig. 2. Schematic Diagram of Compartmental Model

The compartments are so indexed that the reaction $A_{i-1} \rightarrow A_i$ takes place in the ith compartment. Combining rate equations and mass balances, the following set of differential equations is obtained (1):

(4)
$$\frac{dA_{i-1,i}}{dt} = r_{i-1}(A_{i-1,i-1} - A_{i-1,i}) - k_{i-1}A_{i-1,i}A_{i,i} - \sigma_{i-1,i}A_{i-1,i},$$

$$\frac{dA_{i,i}}{dt} = k_i A_{i-1,i} A_{i,i} - r_i(A_{i,i} - A_{i,i+1}) - \sigma_{i,i}A_{i,i}$$

where $r_i = D_i S/\Delta x$ (Δx is the distance between compartments), and

$$\sigma_{i,i} = \frac{h_{i,i} S}{V}.$$

In our work three entry conditions are considered.

a) Precursor concentration is held constant, that is,

$$\frac{dA_{01}}{dt} = 0,$$

(5)
$$\frac{dA_{11}}{dt} = KA_{11} - r_1(A_{11} - A_{12}) - \sigma_{11}A_{11}.$$

b) Precursor enters by diffusion from the surroundings, where its concentration is a constant, W_o, that is,

(6)
$$\frac{dA_{01}}{dt} = r_o (W_o - A_{01}) - k_o A_{01} A_{11},$$

$$\frac{dA_{11}}{dt} = k_o A_{01} A_{11} - r_1 (A_{11} - A_{12}) - \sigma_{11} A_{11}.$$

c) Precursor is introduced by a pseudo first order reaction,

(7)
$$\frac{dA_{01}}{dt} = K A_{01} - k_o A_{01} A_{11}.$$

The two systems were simulated on an analog computer so that a wide variety of parameters covering all possible interactions were studied.

DISCUSSION

For the unstructured or homogeneous model, the way in which precursor enters the cell is critical in determining oscillatory behavior. The two types of entry conditions simulated were precursor entering via diffusion through the cell wall and constant precursor concentration within the cell (infinitely fast diffusion into the cell).

With infinitely fast diffusion across the wall, undamped oscillations are observed for all reacting species. The frequency of oscillations decreases with increasing chain lenght (number of consecutive reaction steps) of reactants, but increases with increasing kinetic rate constants. Since the rate constant is of the form $K = K_o e^{-E/RT}$, it is evident that increasing the cell temperature will serve to increase frequency for the unstructured cell. This dependence is significant when compared with the response of the structured system to increased cell temperature (below). Also of interest for this system is the observation of simple modulation in the wave forms for several cases. With the proper choice of parameters beat frequencies of low order harmonics may be obtained. These are of interest for biological clocks with longer rhythms.

For finite rates of precursor diffusion through the cell wall, the species oscillations become damped, though they persist for long periods of time (decay ratio over 90%) when the permeability of the cell toward precursor σ_o is 50 times the permeability toward an individual specie, σ_s. For precursor species permeability ratios of approximately unity damping is relatively quick. It is interesting to note that the oscillations do not retain their shape as damping increases but become much less structured and regular, so that it is difficult to discern a characteristic frequency. The <u>transition</u> from a damped specie oscillation to the undamped state <u>is a continuous function of permeability ratio</u>. However, the extent of damping is also dependent upon the autocatalytic rate constants of the species as well as the initial precursor concentration.

For the structured or compartmental model the entry condition of precursor again plays a vital role in determining the existence of sustained oscillations. However, the degree of structure of the system (as measured by the ease of with which reacting species can migrate to various compartments) may be the decisive feature determining oscillatory behavior in most cases.

The three <u>types of entry conditions</u> simulated were (1) constant precursor concentration in the fast compartment, denoted by Cl; (2) precursor entering the cell via diffusion (C2); and (3) precursor being formed via a fast order autocatalytic reaction at the first site (C3). Entry condition C3 leads to sustained oscillations in each specie regardless of the magnitude of other parameters in the cell. For this case the equations describing the first compartment are of the VOLTERRA - LOTKA type creating a model oscillation that drives the entire system just as if a sinusoidal input had been applied to the cell. It was noted in these cases that the frequency was approximately the same for all compartments and furthermore changed little for widely varying rate constants, being a function of primarily the first order autocatalytic rate constant for formation of A_{01} (the precursor) (see Eq. (7)) +.

For the homogeneous or unstructured model, the amplitudes of oscillatory species were found to increase with increasing initial disturbance from the steady state. In the case of the C3 type of structure, starting far from the steady state does not give rise to large oscillations. Rather, the species damp down to a limit cycle whose amplitude depends on the system parameters (Fig. 3).

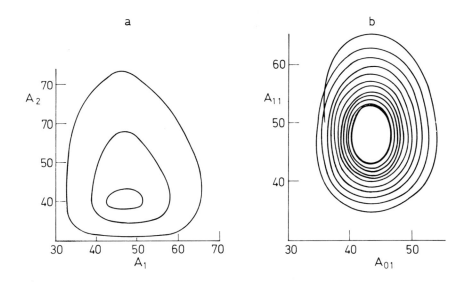

Fig. 3. Phase plane plot (concentration vs concentration) a) homogeneous model b) compartmental model

For entry condition C1, the sufficiency condition of BIERMAN (1) states that r_i is bounded from above and cannot exceed some upper limit without destroying oscillations. However, an examination of the steady equations (neglecting loss to the environment temporarily) reveals that the rate of diffusion between compartments must equal the rate of loss due to reaction, so that increasing the diffusion constant, r_i, tends to make $\overline{A}_{i,i} = \overline{A}_{i,i+1}$; thus, reducing the compartmental system to a homogeneous model. In other words, by increasing the diffusion constant, we are increasing the mobility of species and destroying compartmentalization while preserving the autocatalytic character of the system.

+ However, the amplitude of oscillation decreased for each succeeding compartment down the chain, so that for larger chains even high amplitude fluctuations in the first compartment result in little disturbance farther down the chain. This type of behavior is consistent with the known buffer capacity of cells.

Analog simulation shows that the diffusion is indeed bounded from below. Systems with high ratios of r_i/K_i exhibited damped oscillation while for those with low value of r_i/K_i, oscillations were of low amplitude and long period; thus becoming undetectable. The damping of oscillations thus decreases as the parameter r_i/K_i increases. Damping can be increased either by lowering r_i for a given set of rate constants or by increasing K_i, the rate constant for given diffusivity r_i. Thus, if initially the diffusion constant is large compared with K_i and K_i increases, the damping will likewise increase with an accompanying decrease in frequency. This behavior is contrary to the tendency of the homogeneous cell to oscillate more rapidly with higher K_i. Since an increase in temperature causes a sharp increase in K_i, the oscillations in the compartmental model will be destroyed at higher temperatures. Even more interesting, perhaps, is the phenomenum of suddenly seeing oscillations manifest themselves as the temperature of this system is decreased.

The effect of σ_i, the permeability of intermediate species, is quite pronounced in those C1 systems where oscillations will be obtained with zero specie permeability, hence no loss of intermediate species to the medium. Only small loss to the medium results in a great deal of damping and markedly decrease frequency of specie oscillation. In many cases, it is sufficient to destroy any regular pattern of damped oscillations.(Fig.4.).

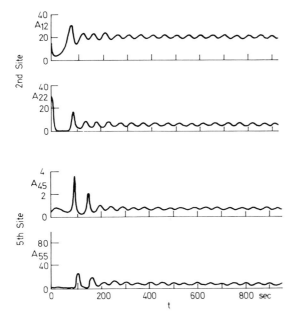

Fig. 4. Effect of specie permeability on oscillations - five site compartmental model (C1)

The system with entry condition C1 in the range r_i/K_i for which oscillations are possible shows similar behavior to system C3 as regards the relationship of amplitude of oscillations versus increasing initial disturbance in that the various species quickly damp down to a limit cycle type of behavior. In the C1 system, the limit cycle will be slightly damped whereas the C3 system shows undamped limit cycle behavior.

A word on initial disturbance may be in order at this point. An initial disturbance may be caused by either a fluctuation in specie concentration or by a fluctuation in temperature. Temperature variations lead to a new set of rate constants in that the steady state of the system is changed, with the result being an apparent shift in relative specie concentration. We have discussed only an initial disturbance although in reality our model system may be subject to many random fluctuations in both temperature and concentration.

One final observation on the structured system with entry condition C1 is worth noting here. The intra-compartmental diffusion constants in the earlier compartments of the chain are critical in determining whether the entire system will oscillate. In a five site case more persistant oscillations were observed with high diffusivities between the first three sites than when the diffusivity magnitudes were reversed. High diffusivity from the first two sites was sufficient to set the entire cell oscillating; whereas high diffusivity at the last two sites oscillations of short duration were localized to the sites directly affected by the high diffusivities.

The models discussed above are, of course, only a highly idealized representation of an interactive-autocatalytic system that may not be physically realizable. However, the simulation uncovers some biologically interesting interactions between diffusion and reaction, in addition to elucidating the interaction between structure, method of precursor entrance, chain length, and temperature fluctuations in initiating oscillations for a complex open system. Hopefully, the kinetic basis for some of the unusual effects described herein can be an aid in interpreting other exotic cellular phenomena.

References

1. BIERMAN, A.: Bull. Math. Biophys. 70, 203 (1954).
2. BRAY, W.C.: J.A.C.S., 43, 1262 (1921).
3. CHANCE, E., ESTABROOK, R., GHOSH, A.: P.N.A.S., 51, 1244 (1964).
4. FINN, R.K., WILSON, R.E.: J. Agric. Food Chem. 2, 66 (1954).
5. FRANK-KAMENETSKII, A.D.: "Diffusion and Heat Exchange in Chemical Kinetics", Princeton Univ. Press (1955).
6. GROSS, B., KIM, Y.G.: Bull. Math. Biophys., 31, 441 (1969).
7. HIGGINS, J.: I & EC, 59, 19 (1967).
8. LOTKA, A.J.: J. Phys. Chem. 14, 271 (1910).
9. MOORE, M.: Trans. Far. Soc. 45, 1098 (1949).
10. WALSH, A.D.: Trans. Far. Soc. 43, 305 (1947).

DISCUSSION

BREMERMANN:
Why were the various rate constants chosen at the values given in the several figures? Are they the result of analog computer experimentation involving trial and error modification of constants? How can these constants be related to the constants involved in the experimental system (such as considered by CHANCE, FRENKEL, et.al).

GROSS:
The constants were obtained by analog experimentation, but frequencies lie well within the actual range reported by others.

WALTER:
BRAY and others have made suggestions, but the source of the oscillations in the iodic acid-catalyzed decomposition of hydrogen peroxide remains unknown. - The most common form of positive feedback in enzyme-catalyzed systems is substrate inhibition. However the equations in this paper do not describe this type of feedback. It is true that positive feedback due to endproduct stimulation is well-known in a very limited number of enzyme systems, and feedback of this type results in kinetics of an auto-catalytic form. It is not true however that "many enzymatic (sic!) reactions are of the autocatalytic form". Since the whole justification for the studies

reported in this paper depends upon the veracity of the sentence on page 65, this sentence should be documented or the context of the equations changed.

GROSS:
1. I meant to say that a plausible mechanism involving autocatalytic reactions which seems to fit the observed data exists.

2. One example of auto catalytic enzyme reactions is the formation of active trypsin from trypsinogen. This paper does not purport to document a specific sequence of reactions, but rather to look at the consequences of one plausible set of reactions that could take place within a cell.

ANDREW:
For the benefit of readers who are new to this area of study, could the authors give some idea of the range of frequencies of oscillation which might be expected in living cells? Some of their figures have time-scales attached, but it is not clear (to me, at least) whether they correspond to systems having parameters approximating those of living cells. Obviously, any attempt to speculate about the part these oscillations might play in living systems requires a rough estimate of the range of frequencies which is possible.

GROSS:
Frequency of oscillations observed in biological reaction can range from 1 sec^{-1} to $\frac{1}{60} \text{ sec}^{-1}$. The frequencies in this paper are within that range. GROSS (In response to Dr. SIMON) + : We have not investigated this facet, and would encourage a more detailed look into optimization and control strategies for these systems.

+ see page 111

Oscillatory Behavior of Enzymic Activities: A New Type of Metabolic Control System

S. Comorosan

Abstract:
Two biochemical types of control units, namely the switching and the two factor dicriminating net are derived as a consequence of the oscillations that can be induced in the enzymic activities by low level UV-irradiated substrates. A complex network encompassing long sequences of metabolic reactions can be constructed, and on this basis the organization of cellular metabolic activities in well defined "regimes" and "states" is inferred.

Reference: Bull. Math. Biophys. 31, 623 (1969).

DISCUSSION

SIMON:
Although the two types of control units proposed by you in order to explain the UV-induced oscillations are interesting I do not find any indication at the reason why the perturbation by low UV-irradiation should produce the observed oscillations.

COMOROSAN:
For the "substrate perturbation phenomenon" it is clear that a "relational" model can predict the kinetic fluctuations, but no more than those. For further clarification of the new experimental effect new experimental data are required.

The Existence of Synchronous States in Populations of Oscillators[+]

Th. Pavlidis

Abstract

A simple analysis illustrates how a continuum of interacting oscillators may also show oscillations in space. This affects the phase of the oscillations in the time domain with the result that two groups of oscillators are formed, oscillating at the same frequency, but with a phase difference equal to half a period. This is called a second order synchronization and it is expected to be less stable and a lower frequency than a first order synchronization in which all units oscillate in phase and with the same frequency. There are indications that certain phenomena associated with circadian rhythms may be due to changes from one type of synchronization into another.

1. Introduction

Although there is an extensive literature on biological and chemical oscillators, very few papers deal with groups of interacting oscillators (6), (13), (20). Part of the reason for the limited amount of work in this area seems to be the mathematical intractibility of higher order oscillatory systems (1), (8), (12). There have been only a few results for systems of two coupled oscillators (7), (12) and there are certain techniques available for the study of special forms of systems consisting of an arbitrary number of interacting units (5), (6), (12), (19). In general, a model of biological oscillators is not expected to meet such conditions. However, one may start with a simplified form which can be analyzed and then complete the investigation by computer simulation. An attempt to go directly into the simulation is not likely to be fruitful because of the complexity of the system and the large number of parameters involved. The approximate analysis will at least give some idea about what to look for. We will follow this approach in this paper by studying first a linear system of oscillators, then indicating what the effects of nonlinearities might be and finally reporting on the results of computer simulation.

We are concerned with the problem of synchronization of oscillators, i.e., a situation where all units oscillate at the same frequency. If all of them are in phase or if the maximum phase difference observed is a small fraction of the period of oscillation we will say that the synchronization is of first order. If they form n groups such that the oscillators within each group are in phase (at least approximately) but there are significant phase differences between groups then we will say that the synchronization is of nth order. We will also use the term <u>a synchronous state (of nth order)</u> to describe the state of the population when synchronization has occurred.

It is obvious that if one observes a population of oscillators macroscopically he will be able to detect the oscillation only when the population has reached a synchronous state. By macroscopic observation we mean in particular one under which there is no way to follow the behavior in individual members of the population. This is the case in a biochemical system whose output is, say, a hormone and one can detect only its total concentration. Or in a population of animals where one observes their circadian activity without being able to label the individual members.

[+] This research was supported by NSF Grant GK-13622.

The eclosion rhythm of insects is certainly such an example (16), (17). Note that under those conditions in an nth order synchronization the observed average frequency will be n times the frequency of the individual units. Thus an experimental observation of, say, a doubling of the frequency might indicate the transition from a first order synchronization to a second order.

There is at least one series of experiments where an approximately twofold increase in frequency has been observed. This is the splitting of circadian rhythms and, in particular, the occurrence of two "peaks" per day in the running wheel activity of a number of both diurnal and nocturnal mammals (10), (16). There are also other experiments which produced results similar to the ones expected from a population of oscillators (e.g. (4)). We hope that a quantitative analysis of the phenomena and the proposed models will help to shed some light into the nature of the mechanisms involved. In particular we hope to find what other properties one should observe if indeed the system consists of a population of oscillators.

2. Approximate Analysis

As a starting point for our analysis we consider a population of oscillators with nonlinear damping terms and with linear elastic coupling. If $q_1, q_2, \ldots, q_n$ are the outputs of each unit, the following set of equations will describe the system:

(1) $$\frac{d^2 q_k}{dt^2} + f(k, q_k) \frac{dq_k}{dt} + \omega_k^2 q_k = \sum_{j=1}^{n} a_{kj} q_j \qquad k=1, 2, \ldots, n$$

If the number of units involved is very large one can describe the system as a "continuum of oscillators". A single variable q will describe all of them, but it will be not only a function of time, but also of another variable, x, which may be thought of as a space coordinate. The variations of q as a function of time at a given point in space can be considered then as describing the output of a single unit. This model may be closer to reality in the case of biochemical systems while the discrete version described by eq. (1) is more realistic in the case of an insect population. If K(x, u) describes the effect of the value of q at u to the value of q at x then the following equation characterizes the system:

(2) $$\frac{\partial^2 q(x,t)}{\partial t^2} + f(x, q(x,t)) \frac{\partial q(x,t)}{\partial t} + \omega^2(x) q(x,t) = \int K(x,u) q(u) du$$

The integral is taken over the space occupied by the system. The limits of integration can be extended trivially to infinity by assuming K to be zero outside the space of interest. Eq. (2) may be a considerable simplification for the dynamics of a biological system, but it is still too complicated for mathematical treatment. Therefore, we proceed with two additional simplifying assumptions.

a) The medium is uniform, i.e., f and ω do not depend on x. This is equivalent to the assumption of identical oscillators in the discrete case and it has been shown to be a reasonable approximation for the case that the variations among the units are small (13).

b) The nonlinear damping term is small and the system will oscillate only at frequencies close to the ones of the undamped linear system. Note that although this assumption is always valid for oscillators with one degree of freedom it is not so for the case of higher order oscillators like the one studied here (1). Therefore, one must be cautious in the interpretation of the results.

With these simplifications eq. (2) reduces to

(3) +
$$\frac{\partial^2 q(x,t)}{\partial t^2} + \omega^2 q(x,t) = \omega^2 \int_{-\infty}^{\infty} K(x,u) q(u) \, du$$

This equation can be solved easily by the technique of variable separation (2), (3), i.e., by expressing q as the product of two functions:

(4) $q(x,t) = X(x) T(t)$

It can then be shown easily that T (t) will be a sinusoidal function with frequency $\omega \sqrt{1-\lambda}$ while X (x) will satisfy the following equation:

(5)
$$\int_{-\infty}^{\infty} K(x,u) X(u) \, du = \lambda X(x)$$

i.e., X (x) is an eigenfunction and λ an eigenvalue of the integral operator described by the left hand side of eq. (5) (3). There will be, in general, an infinity of solutions of eq. (5) and for each one we will have a corresponding solution T (t) and thus q (x, t) as given by eq. (4). A solution of that form will be called a _mode_ of eq. (3). Because of the linearity, the superposition of such solutions will also satisfy eq. (3). Since the frequency of the oscillations depends on λ it is obvious that in order to observe a monofrequency oscillation it is necessary either that only one mode be present or that all the modes present have the same eigenvalues. These are necessary conditions for the existence of synchronous states in terms of our earlier discussion.

If X (x) has the same sign for all values of x then all oscillators will be in phase and this corresponds to first order synchronization. However, if X (x) changes sign then the points corresponding to positive X (x) will have a halfperiod phase difference from the points with negative X (x) and this obviously corresponds to second order synchronization.

The above offers a characterization of the conditions responsible for first and second order synchrony in the system described by eq. (3). We will discuss now briefly two special cases illustrating these situations.

1. Each point is affected only by its neighbors at a distance up to some constant L. It is also affected equally by all of them. i.e.,

$$K(x, u) = \begin{cases} 1 & \text{if } |x-u| < L \\ 0 & \text{otherwise} \end{cases}$$

There is a continuum of eigenfunctions:

a) $X(x) = a + bx$ for any a and b with $\lambda = 2L$
b) $X(x) = a \cos(bx)$ for any a and b with $\lambda = \frac{2}{b} \sin bL$.

2. Each point is affected by all the others, but their effect decreases exponentially with the distance, i.e.,

---

+ The factor ω^2 in the RHS allows us to treat K(x,u) as having dimension (L^{-1}).

$$K(x, u) = \exp\left(-\frac{|x-u|}{L}\right)$$

Again, there is a continuum of eigenfunctions:

a) $X(x) = a\exp(bx)$ for any a and $|b| < 1/L$ with $\lambda = -\dfrac{2L}{1-(bL)^2}$

b) $X(x) = a\cos(bx)$ for any a and b with $\lambda = -\dfrac{2L}{1+(bL)^2}$

In both cases the values of a and b (and also which modes are present) are determined from the initial and boundary conditions. In both cases $X(x) = $ constant is an eigenfunction and this implies that all units have the same amplitude and oscillate at a frequency $\omega_1 = \omega\sqrt{1+2L}$. On the other hand if only one mode of the form $a\cos(bx)$ is present then a second order synchronization will be observed with frequency twice $\omega\sqrt{1-\lambda}$. It is easy to verify that in both cases the latter quantity is always less than ω_1. Thus in physical systems for which the above model is valid one should always expect the observed frequency of the second order synchronization to be less than twice the frequency of first order synchronization.

While in the case of a linear system there exists a large number of possible solutions, a more restricted situation is expected for nonlinear systems. There not all amplitudes of oscillations are possible because the system will exhibit, in general, a limit cycle behavior. If the initial distribution of amplitudes is nonuniform, points with small amplitude might be "inside" the limit cycle while points with large amplitudes might be "outside". Eventually, all of them might converge to the same amplitude. Thus a first order synchronization is more likely for a nonlinear system than a linear. This conclusion is in agreement with earlier studies of similar systems (13), (20). However, identical amplitudes does not always imply synchronization. If the original distribution is of the form $\cos(bx)$ then the process described above may well lead to a distribution of the form $\text{sign}(\cos(bx))$. Then the space will be divided in zones containing oscillators in phase within each one of them but in phase opposition in adjacent zones. The boundaries of these zones are expected to be areas of "stress" and therefore a strongly interacting system will revert to a first order synchronization. However, a weakly interacting system might stay in a state of second order synchronization.

These arguments indicate that a second order synchronization is at least feasible for a nonlinear system although less stable than a first order synchronization. Thus a transition from first to second should be more difficult than the other way around and could occur only in the presence of some outside disturbance. For the case of circadian rhythm splitting, such a disturbance is a change in light intensity. It is well known that this affects the frequency of oscillations and therefore one should expect different periodic trajectories in the state space of the system for different light intensities. The transition between the two trajectories may well cause a scattering of phases so that the new steady state will correspond to a second order synchronization.

3. Results of Simulation

The last discussion is certainly quite qualitative and one may seriously doubt the validity of the conclusions, especially for systems whose mathematical description may be substantially different from eq. (2). For this reason, we proceed with the simulation of a particular system on a computer. Because it is necessary to have discrete variables in such a simulation, the system is effectively one with a finite number of interacting oscillators. As a basic unit we chose a model of a biochemical oscillatory system described by HIGGINS (9). Such a model has also been shown to be able to simulate many aspects of circadian rhythms (15). Each unit receives from its environment a substance G wich is converted into a substrate X. X is transformed into another substrate, Y, in a reaction catalyzed by an enzyme, E_1, according to the MICHAELIS-MENTEN law. Y then is transformed into a third substance, Z, by a similar reaction catalyzed by an enzyme E_2. Y also activates E_1 and this provides a feedback in the system. If one assume that light contributes to the activation of E_2 then the period of oscillations becomes an increasing function of light intensity. The substance Z is now assumed to return to the environment of the system. There the contribution from all the units are thoroughly mixed before interacting with the production of G. This provides a coupling among the units which is additive and with each unit contribution equally to all the rest (in terms of eq. (1) this equivalent to requiring that a_{kj} is identically equal to unity). Such an assumption may not be very realistic because one expects neighboring units to interact in a strong fashion. It also removes the structure in space that the systems discussed earlier possess. However, it was retained because it simplified the mathematical analysis and the simulation of the system. The latter are described in detail elsewhere (14) and we summarize here only the main conclusions of that study:

a) For weak coupling the period of the oscillation is an increasing function of the light intensity. However, for strong coupling this relation is reversed.

b) For a given light intensity in the environment of the units the period is an increasing function of the coupling strenght.

c) For weak coupling it is possible to observe second order synchronization, especially for higher light intensities.

d) If light is assumed to inactivate the enzyme E_2 then the period is a decreasing function of the light intensity for weak coupling and the opposite is true for strong coupling. Also second order synchronization occurs for low light intensities.

These features as well as the underlined conclusions of the previous section seem to agree with experimental results (10), (16) if one assumes that the light activates E_2 in nocturnal organisms and it inactivates it in diurnal organisms. This is an indication that multiple synchronous states may indeed occur in populations of oscillators in nature. A detailed discussion about the implications of this for circadian rhythms is presented elsewhere (14). Here we point out that on the basis of thermodynamical considerations "space oscillations" (of the form described by $X(x) = \cos(bx)$) have been predicted for chemical systems although not in conjunction with oscillations in the time domain (11), (18). Thus one might be justified in thinking that such phenomena should occur in other systems besides circadian clocks. It would be interesting to know if anyone has observed periodic phenomena in a living system which present the salient features of a population of oscillators discussed in this paper.

References

1. BOGOLIUBOV, N.N., MITROPOLSKI, Y.A.: Asymptotic Methods in the Theory of Nonlinear Oscillations, New Delhi: Hindustan Publ.Comp., (1961).
2. CHURCHILL, R.V.: Fourier Series and Boundary Value Problems, New York: McGraw-Hill, 1941.
3. FRIEDMAN, B.: Principles and Techniques of Applied Mathematics, New York: J. Wiley, 1956.
4. GASTON, S., MENAKER, M.: Science, 160, 1125, (1968).
5. GELB, A., VANDER VELDE, W.E.: Multiple-Input Describing Functions and Nonlinear System Design, New York: McGraw-Hill, 1968.
6. GOODWIN, B.: Temporal Organization in Cells, New York: Academic Press, 1963.
7. HAAG, J.: Oscillatory Motions, Belmont, California: Wadsworth Pub. Comp. 1962.
8. HALE, J.K.: Oscillations in Nonlinear Systems, New York: McGraw-Hill, 1963.

9. HIGGINS, J.: Ind. Engin. Chem. 59, 19 (1967).
10. HOFFMAN, K.: Biochronometry Symposium Proc., Ed.: Menaker, M. (in press).
11. LEFEVER, R., J. Chem. Phys., 49, 4977 (1968).
12. MINORSKY, N.: Nonlinear Oscillations, New York: Van Nostrand 1962.
13. PAVLIDIS, T.: J. Theor. Biol. 22, 418 (1969).
14. PAVLIDIS, T.: J. Theor. Biol. 33, 319 (1971).
15. PAVLIDIS, T., KAUZMANN, W.: Arch. Biochem. Biophys. 132, 338 (1969).
16. PITTENDRIGH, C.S.: Cold Spring Harbor Symp. Quant. Biol. 25, 159 (1961).
17. PITTENDRIGH, C.S.: Z. Pflanzenphysiol. 54, 275 (1966).
18. PRIGOGINE, I. et al.: Nature 223, 913 (1969).
19. SANDBERG, I.W.: Bell System Techn. Journ. 48, 1999 (1969).
20. WINFREE, A.T.: J. Theor. Biol. 16, 15 (1967).

DISCUSSION

BREMERMANN:
What about heart contractions? Heart tissue is a population of synchronous coupled non-linear oscillators. When synchrony breaks down fibrillation occurs. There are interesting phenomena that depend upon the size of contiguous tissue. In other words small pieces of heart tissue behave differently from sufficiently large pieces. - There has been quite a bit of interest in these questions in the biomathematics group around I. GELFAND and FOMIN at Moscow.

PAVLIDIS:
I am not familiar with the types of different behavior of heart tissue. I can think of one situation where the size of the piece of tissue will affect frequency of contractions. This can be achieved by setting the coupling coefficients of eq. (1) equal to each other and to a fixed percentage r of $\omega^2 (= \omega_1^2 = \ldots = \omega_k^2)$:

$$a_{ki} = r\omega^2 \text{ for all } k, i$$

Then it can be shown (Ref. 13) that the frequency of the stable synchronous oscillation is given by

$$\omega_1 = \omega\sqrt{1 + r(n-1)}$$

In this case the size of the population affects the frequency directly. This phenomenon occurs even if the characteristic frequencies are different and the coupling asymmetrical, provided the differences are not very large (Ref. 13).

ROSEN:
Some years ago, ULAM presented some results (which as I recall were only incompletely published) on simulations he had carried out concerning populations of nonlinear mechanical oscillators. He claimed that the simulations revealed certain unexpected results, in particular with regard to the partitioning of energy through the various oscillatory modes available to the system. It seems to me that such results might be made to follow from the kind of analysis given herein, and would be of value for an understanding of the kinds of modes of oscillations favored in particular populations (and also for the probabilities of switching between such modes).

PAVLIDIS:
Modes corresponding to lower order synchrony seem to be favored over those of higher order,

provided the mutual coupling is strong enough (see my comments on Dr. FONG's question). A related observation is the one by HAYASHI, that for single oscillators, externally driven, subharmonic synchronization is more difficult to be attained than harmonic synchronization. There seems to be a relation between subharmonic entrainment and higher order synchronization in populations of oscillators (Ref. (14)).

COULTER:
I am interested in the possible applicability of Dr. PAVLIDIS' theory to synchronization of neural oscillators, especially those involved in brain rhythms. For these, more than one spatial dimension may be necessary. Has PAVLIDIS considered a generalization to two, three, or n spatial dimensions, and if so, how would this affect the necessary conditions for the existence of synchronous states?

PAVLIDIS:
The method is readily extendable to n spatial dimensions by simply interpreting the variables x and u in eq. (5) as vectors and the integral as a volume integral. However determining all the eigenfunctions in this case might be considerably more difficult.

A simple example for two dimensions can be obtained by defining

$$K(x, u) = \begin{cases} 1 & \text{of } |x_1 - u_1| < L \text{ and } |x_2 - u_2| < L \\ 0 & \text{otherwise} \end{cases}$$

One possible eigenfunction is the following:

$$x(x) = (a+bx_1) \cos cx_2$$

It would be interesting to determine Kernels which have eigenfunctions whose maxima lie on specific geometric curves, e.g., concentring circles or spirales. This could explain the patterns appearing in fungi cultures (BOURREL et al. in Science, 166, 763, 1969) and would offer a quantitative model of the phenomenon extending the one described by WINFREE (1970 IEEE Symp. on Ad. Proc. XXIII, 41).

Regarding neural oscillators I think that there exist additional difficulties because of the combination of continuous and discrete processes (the firing of pulses). The study of a single oscillator for example is already quite involved (PAVLIDIS, Bull. Math. Biophys. 27, 215, (1965), and RASHEVSKY, Bull. Math. Biophys. 33, 281 (1971)).

FONG:
The solutions representing synchronous states are a few out of infinitely many which are presumably not synchronous, e.g., solutions of the form Eq. (5) $X(x) = \xi(x) e^{i \eta(x)}$ where $\xi(x)$ and $\eta(x)$ are real functions of x. $\eta(x)$ represents the variation of the phase with respect to x, signifying non-synchronousness. To be biologically significant it seems necessary to show that either these non-synchronous states do not exist or there are special conditions that make the synchronous states much more probable to occur. In the absence of such demonstrations we are still faced with the question why some biological systems favor some special solutions out of infinitely many possible solutions. – The method of solving a set of ordinary differential equations by substituting for them a single partial differential equation is a powerful one and can be applied quite generally. In fact I myself have used this method a few years ago to solve a problem concerning the origin of chemical elements in nucleogenesis in stars with remarkable success (Physical Review, 120, 1388 (1960)) – the partial differential equation was analytically solvable.

PAVLIDIS:
As it can be seen from the examples I consider solutions of the form $\xi(x)e^{i\eta(x)}$ as synchronous (of higher order). Loss of synchrony occurs only when more than one eigenfunctions (with corresponding different temporal frequencies) are present. This in turn depends on initial, boundary conditions and choice of parameters. The fact that groups of interacting oscillators tend to mutually synchronize is a direct consequence of the properties of nonlinear entertainment and it has been discussed by a number of authors including GOODWIN (6), WINFREE (20) and myself (13),(14). This feature is true for more general systems and not just biological. My present discussion was motivated by some very striking experiments by PITTENDRIGH (Ref. (16) and in Life Science and Space Research, 122, 1967) and HOFFMAN (10).

LOCKER:
Although your paper offers an explanation for the synchronization of oscillatory activities, I would like to know how in your opinion an oscillatory state originates and what kinds of oscillatory mechanisms could have played a role in biogenesis. What are the means by which weak or strong interactions between single oscillators or oscillatory states could be brought about? Your demonstration of the connexion between order of synchronization and stability could be applied to a hypothesis of the emergence of bio-systems which, of course, in a certain sense are the more labile the more organizational complexity, e.g. synchronization or entrainment of the activity of sub-systems, they exhibit.

PAVLIDIS:
I will try to outline a conjecture about the biogenesis of oscillatroy systems: Feedback control has a definite survival value since it allows an organism to compensate for environmental changes and preserve rather constant interval conditions (homeostasis). However feedback systems quite often present oscillations in response to external disturbances. This is a general property of dynamical systems, not only biological ones. Depending on the nature of the feedback loop and the controlled system, such oscillations may be sustained. Thus it is possible that the first biological "clock" came as a result of "imperfections" in a feedback regulator. If the period of such oscillators was close to an external period (e.g. 24 hours) then their utilization to control behavior contributes to the survival of the organism. The existence of populations of oscillators can be thought of as a direct concequence of the structure of the organisms.

Reactions of Model-Oscillations to External Stimuli Depending on the Type of Oscillation

R. Wever

Abstract +
One of the fundamental principles in biology is the rhythmicity which is characterized by the presence of self-sustained oscillations at very different levels of organization. In order to systematize our knowledge of living systems, a classification is necessary, even with regard to the properties of biological oscillations. It is the end of this paper to find out empirical criteria which allow to classify oscillations in one distinct respect, and that to discriminate between pendulum oscillations (with a small energy exchange between the oscillating system and the environment) and relaxation oscillations (with a large energy exchange), by means of model simulations. Biological references are taken from circadian rhythms as the best known biological oscillations.

The most important criterion is based on the variability of successive periods within an oscillation, either as a consequence of random noise, or of periodic perturbations being not able to synchronize the oscillation. This criterion states that the ratio between the variability coefficient of frequency (or of period) and that of amplitude, in each case formally calculated from the parameters of successive periods, is positively correlated (or even proportional) to the energy exchange characterizing the oscillation type; this ratio is smaller than one in pendulum oscillations, and it is larger than one in relaxation oscillations.

Other criteria are applicable to oscillations when entrained by Zeitgebers. For examining this, the existence of two borderline cases of entrainment must be taken into account. In both cases, amplitude as well as frequency of the oscillation vary periodically; in 'relative coordination', the oscillation is, in the average, not entrained; in contrast to this, in 'relative entrainment', the oscillation is entrained but only in the average. Bearing this in mind, the range of entrainment, under the influence of a definite Zeitgeber, is smaller in pendulum than in relaxation oscillations. Moreover, under the influence of a strong Zeitgeber, the transition between the primary range of entrainment (with 1:1 synchronization) and the secondary one (with 1:2 synchronization) is continuous in pendulum oscillations but discontinuous in relaxation oscillations. This means that, with increasing Zeitgeber period, the unimodal course of the oscillation changes steadily, without any discrete limit, into a bimodal one in pendulum oscillations, but with a sudden transition in relaxation oscillations.

DISCUSSION

PAVLIDIS:
There are two comments which I would like to make on this paper: a) The computer simulation of the VAN DER POL oscillator shows that the range of entrainment is an increasing function of both the amplitude of the Zeitgeber the value of ϵ. It is possible to derive some related results theoretically using a method originally proposed by ANDRONOV and WITT in 1930. MINORSKI gives a detailed account of it (' Nonlinear Oscillators', Van Nostrand, 1962, pp. 438-444).

+ Full paper appears presumably in BG&Th.

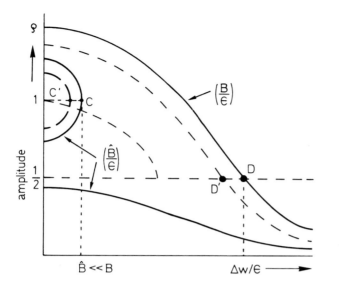

Fig. 1

Fig. 1 is adapted from that text and shows the normalized amplitude of the oscillation as a function of difference in frequency Δw with the amplitude of the Zeitgeber B as a parameter. The shaded areas correspond to unstable states which are not observable experimentally. For a given, large, ratio of B/ϵ the intersection of the corresponding curve with the shaded region boundary gives the limits of entrainment (point D). For a small ratio of $\hat{B}/\epsilon$ on the other hand, the rightmost tangent on the corresponding curve gives the limits of entrainment (point C). It is obvious that the range of entrainment is an increasing function of B. The effects of changes in ϵ are less obvious: One can plot a relation $Y = f(x)$ between the variables $B+/\epsilon \equiv y$ and $\Delta w+/\epsilon \equiv x$ corresponding to the limits of entrainment from the coordinates of points like C and D in Fig. 1. Clearly $f(0) = 0$. Then a simple argument shows that if $f(x_2)/f(x_1)$ is less than x_2/x_1 then indeed the range of entrainment will be an increasing function of ϵ. It would be an interesting project to complete this analysis and compare its results with the computer simulation. b) The phenomenon of "relative coordination" is discussed briefly by MINORSKI as one of almost periodic oscillations (ibid). It is not surprising that in this case one observes regular phase response curves because there is still a free-running oscillation. In contrast to this in the case of "relative entrainment" there is no free-running oscillation and therefore the absence of phase response curves is to be expected. A detailed discussion of the "oscillatory free-runs" as a frequency modulation phenomenon is given in a paper by me ("Amer. Natural." 103, 31-42, 1969). "Relative entrainment" could be described as amplitude modulation.

WEVER:

a) It is of great interest that the dependency of the range of entrainment, at least with weak Zeitgebers, on the amplitude of the Zeitgeber, and on the value of ϵ, can be derived theoretically. I agree that it would be an interesting project to complete the theoretical analysis mentioned. Especially, in comparison to the computer simulation, such a completion would be advantageous, with weak Zeitgebers, in regard to the various borderline cases of entrainment, and, with strong Zeitgebers, in regard to the transitions between the ranges of entrainment of different orders. Both these topics are of greater relevance referring to the type of oscillation than the size of the range of entrainment.

b) In the borderline cases of entrainment mentioned, in 'relative coordination' as well as in 'relevate entrainment', both, amplitude and frequency of the oscillation vary periodically; it is true, in relaxation oscillations, the variability of amplitude is small, but it is significant. For this reason, it is impossible to describe the one case as frequency modulation (better: phase modulation), and the other one as amplitude modulation. Another reason is the absence of any modulating oscillation. "Relative entrainment" is similar to a beat phenomenon; it contains the free-running oscillation as well as the Zeitgeber oscillation, as can be shown by a frequency analysis. Therefore, the statement of PAVLIDIS that there is, in case of "relative entrainment", no free-running oscillation, is inexact. Moreover, his conclusion that therefore the absence of phase response curves is to be expected, seems to be not convincing: On the one hand, there is a phase response curve, although not in the familiar shape; on the other hand, even in the absolute entrained state, and this means, in the absence of any free-running oscillation, a response curve must be present, because most hypotheses basing on response curves, deal with the derivation of phase-angle differences in the entrained state by using response curves.

Abiogenic Aspects of Biological Excitability. A General Theory for Evolution

N. W. Gabel

Abstract

Excitability, as a property of matter, is examined within the context of information theory and evolution. Excitability is the property of an organized group of particles to transfer environmental information along its parts resulting in the maintenance of its structural integrity. Structuralization is differentiated from organization and explicitly defined as organizational schemata endued with survival potential. As a corollary, value can be defined as the information parameter which expresses potential survival benefit to the receiving particle matrix. The evolution of matter is followed up to the level of a protolife excitable membrane.

Excitability and Information

A biologically excitable tissue is one in which changes in cellular membrane potential (EMF) are transmitted in a perpendicular direction with respect to the potential gradient of the membrane. Implied within this definition is the provision that the velocity of these transmitted potential changes must be measurable. The two animal tissues which exemplify this condition are neural tissue and muscle. The survival benefit of excitability to a biological system (a structuralized particle matrix) is the facilitation of environmental information transfer. During the course of biological evolution neural tissue apparently becomes specialized in the transmittance of information which results in the cephalization of multi-organ systems. This tendency towards cephalization is demonstrated among invertebrates as well as vertebrates indicating some impetus towards increased development of excitability as evolution proceeds.

If excitability were responsible for evolution in the direction of more highly structuralized matrices, then it should be capable of general application to the evolution of matter. If, on the other hand, the property of excitability is responsible for biological evolution alone, we are then left with the doctrine of vitalism which completely abrogates the whole question of the origin of life. It also can be argued that the excitability of inanimate particles is not related to the excitability of living particles. Since both usages of the word excitability are derived from phenomenic observation, the distinction of inanimate excitability from animate excitability must be due to the observers.

The relationship of excitability and evolution was proposed as a hypothesis in 1965 (4). Within the framework of this hypothesis, excitability was defined purposefully with a view towards general applicability, although this general applicability was not stressed. Since the hypothesis dealt specifically with precellular organization, only those aspects which contributed to the formation of the protocell were considered.

Since the value of any hypothesis lies in its applicability, the most expedient thing to do is to state the hypothesis and examine it. For the purpose of definition, excitability is the property of an organized group of particles to transfer environmental information along its parts resulting in the maintenance of its structural integrity (4). Structuralization should be differentiated from organization and explicitly defined as organizational schemata endued with survival potential.

If we assume that information is necessary to maintain organization, then survival efficacy will be enhanced in those structures which best utilize impinging information. Information impinges on structural matrices in the form of mass and energy. If this information were not transferred and utilized, the structural organization would disintegrate because it is the information, in itself, which is the disruptive reaction. It should be noted, however, that information, in the theoretic sense, has been associated with lack of organization (2). This has led to the unfortunate conclusion that increased structuralization as expressed by evolution is contrary to the universal increase of entropy with time (e.g. (3), (22)). WILSON (25) has disputed this conclusion and maintains that an increase in BRILLOUIN's bound information should be equated with an increase in entropy. NEWMAN (14) maintains that the position which identifies increased organization with more information (in the theoretic sense) fails to distinguish between the information available for use within a system and the information that is required to characterize a system. The confusion seems to have arisen from (1) BRILLOUIN 's failure to define value (26), (2) the erroneous equating of structuralization with organization, and (3) an oversight committed in defining system boundaries. As a corollary of the definition of excitability, BRILLOUIN' s value can be defined as the information parameter which expresses potential survival benefit to the receiving particle matrix. Fig. 1 has been constructed to depict the boundaries of an evolving system.

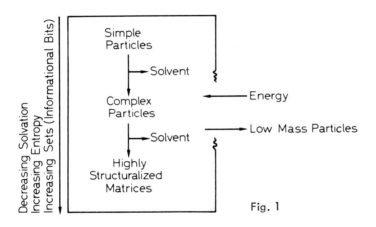

Fig. 1

Perfect randomness and equilibrium does not exist as a geophysical reality. If non-structuralized particles within a system are considered to be quasi-ordered, then the nucleation of a structuralized matrix will result in a greater state of disorder for those particles which do not become part of the matrix. These disordered particles, as informational bits, remain within the system boundaries for a finite lenght of time and exhibit greater degrees of freedom. This phenomenon is profoundly demonstrated by the spontaneous formation of the highly structuralized coordination complexes of solution chemistry. Fig. 1 was generalized from the chemistry of solutions to fit other particulate systems. To pass from this generalization to specific examples, it is only necessary to examine simple structures.

Relaxation

Relaxation can be defined as the lag of response of a system to a change in the forces to which it is being subjected (23). In this context, excitation could be defined as the period (energetically opposite in sign) prior to the state preceding relaxation. This lag is a very general phenomenon and its mechanism and rate depend upon the nature of the forces as well as upon the system. The relaxation energy is that amount of energy which a system either emits to or absorbs from beyond its boundaries during relaxation.

1. Neutron scattering. Since the neutron is an uncharged particle, its coulombic repulsion as it approaches a nucleus is nil, even if its energy is very low. The interacting of a neutron and a nucleus can be represented by the intermediate formation of a compound nucleus which may then react in several ways (21). If the neutron is released, with the reformation of the original nucleus, the process is called scattering. The approach is facilitated if the neutron energy approximates an unoccupied nuclear energy level. A resonance or excitation occurs which facilitates capture of the neutron. This excitation would seem to imply an awareness on the part of the nucleus of the approaching neutron. The impinging information in this case would be the mass-energy parameters of the neutron. The time during which this compound nucleus would exist, would be a nuclear excited state. During this time the nucleus could either rupture, incorporate, or release the captured neutron. Release of the captured neutron would be followed by relaxation.

2. Absorption and emission of electromagnetic radiation. A more vivid example is given by phosphorescent substances (18). If the irradiation of a phosphorescent substance is suddenly stopped, the phosphorescence does not cease immediately, but continues for some time during which the intensity decreases exponentially. The metastable state to which some of the molecules are transferred during irradiation causes the appearance of phosphorescence and is characteristic of the molecules themselves. The molecules or crystalline matrix could be described as having a predisposition (in the same sense as pathological phenomena) to selected radiation (information) which is incorporated by virtue of the capability of some of the molecules to be transferred to a metastable state. Several mechanisms exist for the rapid transfer and dissipation (utilization) of this excitation energy (information) within a given molecule or crystalline matrix. The survival of the particle matrix is dependent, in this case, on the rapid transmittance of information to a metastable state and eventual emission as phosphorescence during relaxation.

Structuralization

Evolution, as opposed to retrogression, is the result of disruption and recombination to more highly structuralized matrices. Following the impact of Darwinism on biology, evolution was considered for some time to be characteristic of living systems only. Research into social evolution was actually interrupted (12). With the advent of successful experiments (13), (1) it became apparent that primordial evolution leading to biochemically important compounds could also proceed on a simple chemical level (15), (17).

Since organized matrices are the product of collisions of energetically excited particles, maximum structuralization requires a high ratio of quantity of energy to number of particles with the further provision that the relatively less thermodynamically stable aggregates do not extensively collide with each other when energetically excited. This condition for the evolution of matter is met by geophysical reality and is borne out by the results of laboratory experiments in chemical evolution (6). For the purpose of illustration, two examples of structuralization analogous to those discussed as relaxation phenomena are given forthwith herein.

1. Neutron capture. If a neutron is retained within a compound nucleus, even though there may be a subsequent disruption into other products, the process is referred to as capture or absorption. If an unstable compound nucleus is formed, disruption can lead to the formation of smaller nuclear fragments (which may or may not reorganize) or to the emission of low-mass, high-energy radiation. The information would again be the mass-energy parameters of the approaching neutron and the possibility of achieving stabilization via the aforementioned emissions could result in a higher degree of structuralization.

2. __Absorption of electromagnetic radiation__. If the impinging radiation to which a molecule is subjected is sufficient to initiate bond-breaking, the fragments may recombine to form more highly structuralized matrices. Since the probability of escape from a gravitational or electrostatic field is maximized in particles having the highest energy to mass ratio, it follows that the escape of simple, less structuralized particles from the reaction site will be facilitated leading to the eventual formation of more complex molecular structures. The profusion of laboratory experiments which have been conducted for the purpose of producing more complex materials from simpler chemicals by radiolysis has been discussed in terms of their relevance to chemical evolution (6). The escape of low-mass particles from a system (immediate environment) appears to be a prerequisite for increased structuralization. Systems which are in equilibrium do not evolve.

The radiation parameters, rather than the absolute quantity of radiant energy, is the information since only energy with wavelenghts which can excite the bonds of the initial molecules are incorporated and utilized. If the residue of high-energy, low-mass particles which are released as a result of greater structuralization are still considered to be part of the total system, then this system would be acting in accordance with the laws of equilibrium thermodynamics.

Precellular Organization

One of the unresolved problems of morphological models for precellular organization is the importance of response to environmental stimulus on evolution (16). Although OPARIN's (15) coacervate model would be responsive to its aqueous environment, its ability to maintain its structural integrity while being impinged by environmental information would be dependent on the composition (material and spatial) of the coacervate drop. Maintenance of structural integrity alone is not evolution. For evolution to occur, the morphological model must in some way utilize the impinging information to increase its responsiveness and impart stability to the __principles__ which govern its structural __pattern__.

DARWIN's theory of natural selection places great stress on survival benefits. Since contemporary life is a metastable, dynamic, equilibrium state, it also is subject to the same general principles which underlie excitability; i.e., incorporation, transference, and utilization of environmental information. The preceding examples of natural phenomena were discussed in order to illustrate the general applicability of the definition of excitability as it was used in its relationship to evolution. In contemporary organisms, phosphates, both organic and inorganic, and the metal ions calcium, magnesium, sodium, and potassium are known to play a dominant role in the transmission of an electrical impulse along an excitable membrane. The transferring of environmental information within a creature is a fundamental life process which would necessarily have been present during the evolutionary processes leading from arbitrarily designated non-living entities to living forms. It is highly unlikely that the chemistry of this method of information transfer would differ markedly in contemporary organisms from the chemistry of primordial forms. A chemical change which would have been disruptive to the structuralization of protolife probably would have involved the solvent environment. Through evolution structural sophistication of excitable tissues can be expected to have taken place, but the fundamental chemical processes probably would remain the same within the axons. Structural evolution, however, would present the possibility for differentiation to occur in synaptic transmission, e.g., electrical versus chemical synapses.

The chemical similarity of the excitability of biological materials to the excitability of polyphosphate complexes induced by exterior variations in the solvent has been discussed in detail (4). It is a characteristic property of inorganic polyphosphates to form coordination complexes with metal cations in aqueous solutions (10), (24). The formation of these metastable structures proceeds with a decrease in free energy of the system and, concomitantly, with an increase in entropy of the system due to the release (to the system) of solvent molecules previously bound to the individual parts of the macrostructure prior to their organization.

The occurrence of inorganic polyphosphates in a large number of microorganisms is well documented (8), (11). It now appears that polyphosphates are also present as cellular constituents of vertebrate tissue and may have a polymodal functionality (7). Neural phosphorproteins which are rapidly labeled by ^{32}P - orthophosphate and which have been postulated to play a role in neuronal transmission (9) may be polyphosphateprotein complexes (7). It has been reported recently, in experiments which attempt to simulate primordial conditions, that polyphosphates could have been present before the emergence of recognizable life-forms (5), (19), (20).

Raison d' être?

The substances comprising excitable tissues have the same properties which they would have in a non-living system. What makes them part of a living system is their structuralization and relationship to other subsystems of a creature. Life did not acquire excitability but was excitable by the very nature of its organization. Excitability could be evolution's raison d' être. If consciousness parallels the development of cephalization and is in reality an expression of excitability which has culminated in human reflection, then not only is the animistic viewpoint of life in nature credible but also the converse viewpoint of life being a composite of non-living interrelationships. The use of anthropomorphic terminology to describe the behavior of both fundamental particles and complex biological entities is not a semantic accident but is the direct result of phenomenic observation which is the basis of operational philosophy. It would be incongruous within the analytic method of phenomenic observation to maintain that terminology describing the behavior or particles and the behavior of complex living organisms are not related. What separates them is the scientists' fear of being accused of mysticism and the aversion of some vitalists to the presumed debasement of life. Life has been delimited from non-life in the same manner and for the same reasons that academic disciplines are delimited from each other. This fragmentation of academic disciplines has resulted in the limitation of much of scientific cognition to a series of unrelated hypotheses. Perhaps it would be preferable in attempting to understand the origin of life to either set no limits or to set new limits.

References

1. ABELSON, P.H.: Paleobiochemistry, Carnegie Institute, Washington Yearbook, 53, 97 (1953 - 1954).
2. BRILLOUIN, L.: Science and Information Theory, 2nd. ed., New York: Academic Press 1962.
3. DANCOFF, S.M., QUASTLER, H.: In: Information Theory in Biology, (QUASTLER, H., ed.) p. 263. Urbana: University of Illinois Press 1953.
4. GABEL, N.W.: Life Sci. 4, 2085 (1965).
5. GABEL, N.W.: Nature, 218, 354 (1968).
6. GABEL, N.W., PONNAMPERUMA, C.: In: Exobiology. (C. PONNAMPERUMA, ed.) Chapter 4. Amsterdam: North-Holland Publishing Company 1971.
7. GABEL, N.W., THOMAS, V.: J. Neurochem. 18, 1229 (1971).
8. HAROLD, F.M.: Bacteriol. Revs., 30, 772 (1966).
9. HEALD, P.J.: Phosphorus Metabolism of Brain, New York: Pergamon Press 1960.
10. JOHNSON, R.D., CALLIS, C.F.: In: The Chemistry of the Coordination Compounds. (BAILER, J.C., Jr. ed.) Chapter 23. New York: Reinhold Publishing Corp. 1956.
11. KULAEV, I.S.: In: Molecular Evolution, Vol. I, Chemical Evolution and the Origin of Life, (BUVET, R., PONNAMPERUMA, C. eds) Amsterdam: North-Holland Publishing Company, pp. 458 - 465, 1971.
12. LEIBOWITZ, L.: J. Theoret. Biol. 25, 255 (1969).

13. MILLER, S.L.: Science, 117, 528 (1953).
14. NEWMAN, S.A.: J. Theoret. Biol. 28, 411 (1970).
15. OPARIN, A.I.: The Chemical Origin of Life (Trans. by A. Synge) Springfield, Ill.: Charles C. Thomas (1964).
16. PONNAMPERUMA, C., CAREN, L., GABEL, N.W.: In: Cell Differentiation (SCHJEIDE, O.A., DE VELLIS, J., eds.) p. 15, New York: Van Nostrand Reinhold Company 1970.
17. PONNAMPERUMA, C., GABEL, N.W.: Space Life Sciences, 1, 64 (1968).
18. PRINGSHEIM, P.: Fluorescence and Phosphorescence, p. 285. New York: Interscience Publishers 1949.
19. RABINOWITZ, J., CHANG, S., PONNAMPERUMA, C.: Nature 218, 442 (1968).
20. RABINOWITZ, J., WOELLER, F., FLORES, J., KREBSBACH, R.: Nature 224, 796 (1969).
21. RAINWATER, L.J., HAVENS, W.W., WU, C.S., DUNNING, J.R.: Physical Revs. 71, 65 (1947).
22. SCHRÖDINGER, E.: What Is Life? Cambridge: University Press 1944.
23. SMYTH, C.P.: In: Molecular Relaxation Processes, Chem. Soc. Spec. Publ. No. 20, London and New York: Academic Press 1966.
24. VAN WAZER, J.R., CAMPANELLA, D.A.: J. Am. Chem. Soc. 72, 655 (1950).
25. WILSON, J.A.: Nature 219, 534 (1968 a).
26. WILSON, J.A.: Nature 219, 535 (1968 b).

DISCUSSION

ROSEN:
"Excitability" is an example of a functional term, not directly susceptible to a general quantitative definition, but only definable ostensibly. For instance, we can take a particular physical system, and mandate that a particular response to a particular environmental forcing will be called an "excitation" (this much is quantitative), and then say that any sufficiently similar response by any other system will also be called "excitable". Whether any particular kind of response is then labeled "excitability" or not is a matter of judgment.

"Excitability" in this loose sense does not seem to be selected for by itself, but only as one aspect integrated as part of a larger organization. In anthropomorphic terms, it seems not an end in itself, but rather a means to an end. Indeed, excitabilities in biology seem generally to come in pairs, so that any particular biological process is a result of an interaction between the simultaneous excitation of a particular mechanism and another mechanism, antagonistic to the first.

GABEL:
I must admit that my definition of excitability is observational rather than operational. Experimentally, however, excitability is always defined operationally whether the observer happens to be an electrophysiologist or a nuclear chemist. The similarity in the behavior of populations of "things", whether they are parts of inanimate structures, biological entities, or social institutions, seems, however, to be too striking to ignore. Therefore, experientially, I felt justified in generalizing a definition of excitability based upon observation.

I did not mean to imply, nor do I really believe that I did, that excitability "is an end in itself". We would seem to be in complete agreement that it is "a means to an end". The end being, quite obviously, evolution.

BRUNNGRABER:
In the text of your paper you define structuralization as a state, but in Fig. 1 structuralization is shown as a process. Could you clarify this point?

GABEL:
You found an oversight in my submitted manuscript for which I must thank you. In generalizing this theory I wasn't quite sure whether it should be presented as a "transformation" or "emergence" theory (ROSEN, This symposium). Hence, the discrepancy between the text and the figure legend. Actually, any process becomes a state if time becomes a constant rather than a variable parameter. Can the theory presented be described as purely emergence or transformation? Perhaps, Dr. ROSEN will supply the answer. It is not uncommon for evolutionary phenomena to defy categorization of this type. For example, society did not spring from any individual human being or group of human beings. It is the convergence of the individual units of humanity upon themselves which have created society. Is this emergence or transformation? Perhaps, theories which can be described as purely "emergence" or "transformation" theories are not applicable to the origin of hierarchies.

LOCKER:
Excitability is in your opinion something like informability as such, namely the property of living organisms to receive information (in its broadest informal sense) and to behave in accord with that, e.g. to alter its state of structuralization. The transfer of environmental information you insert in your definition is somewhat less general since such is only possible after information reception. It is hard to understand why information, if not transferred (and so-to-speak locally absorbed) would disintegrate the structural organization. It is so, in your opinion, because we are dealing with bound information according to WILSON. On the other hand, even the distinction between information used by the system for its own work and the information the observer uses to characterize the system, does not allow one to set information equal with entropy. A way out of the dilemma is to systems-theoretically conceive of the interaction between system and environment. It has been shown (YOCKEY, 1958; ATLAN, 1968) that under the influence of external stimuli an increase of the information content can be brought about. In continuing this line of argument, we would say: This increase cannot be made indefinitely, but rather so long as the system is able to adjust (or even improve) its organization. If the optimally possible state has been surpassed the further increase of information content would become detrimental. Exactly, now that part of information comes into play which cannot advantageously be used by the system, but is possibly of value for the observer to note and to assess e.g. the injurious state of the system. Let us, therefore, assume the centrality of the position of the system and we could reconcile the two seemingly opposing conceptions of information, as exemplified by BRILLOUIN and WILSON, respectively. I am pleased with your idea that life is phenomenologically distinguishable from non-living entities and must not, as I would interpret your statement, quasi-scientifically be derived from an explanation that in its very nature is only physical.

GABEL:
The difficulty which you encounter in understanding why I inserted transference and utilization of information in my definition of excitability would seem to me to be absolved further on in your commentary by your reference to the work of YOCKEY (1958) and ATLAN (1968). We are speaking of the same phenomenon. There doesn't seem to be, in my mind, any disagreement. I really did not equate information with entropy. I merely stated that they both increase in the same direction. Optimization in evolution always falls short of 100 per cent efficiency. When any system reaches the point where its **structuralization** (my definition) becomes endangered, even further organization (systematic arrangement) in the direction of 100 per cent efficiency will eventually be interrupted by disruption. This concept of optimization has been eloquently described by SMITH (1970) + in his article on research management. - I am delighted with your proposal to assume the centrality of the position of the system, since I believe that much of the confusion in the area of evolution has resulted from egocentricity. You are correct in assuming that I am of the opinion that living and non-living entities can be distinguishable only on a phenomenological basis.

+ SMITH, W.V.: Science 167, 957 (1970)

Thermodynamic and Statistical Theory of Life: An Outline
P. Fong

Abstract
Life is regarded as made of three basic elements: matter, energy and information, the basic laws of which are all present in physical science. In particular the laws of creation and dissipation of information are given by the theory of fluctuation and the second law of thermodynamics respectively. On the basis of this foundation we derive from the basic laws a set of secondary principles dealing with the safe-keeping of information, the accumulation of information, the change of information content, the evolution and the revolution of information systems. These principles are employed to explain the origination and evolution of life and to find explanations of peculiar characteristics of like including reproduction, growth, aging, death, sex, choice, control, purpose, ontogenesis, subconciousness and so on. The same may be applied to other information systems such as the mind, the society and cultural developments. Instead of trying to understand the nature of these systems by proposing the a priori existence of an assumed information content we regard any system to begin with zero information content and analyse the origination, selection, accumulation of information and the evolution of the system leading to the present state by the principles developed; we find at the same time explanation of the nature of these systems.

This paper is a brief outline of a theory that tries to establish theoretical biology on the model of theoretical physics upon the foundations of the existing physical science which is generally regarded as closed, complete and exhaustive for the low energy phenomena. A more detailed account will be published elsewhere (1). The work follows a previously published paper entitled Phenomenological Theory of Life (2) , which will be referred here as Paper I; the two together provide an account of the physical basis of life.

1. The Three Elements of Life

Our basic concept is that life is made of three basic elements: matter, energy and information. Information is defined as any spatial, temporal arrangement or relation of entities that is other than random. A system that has an information content is called an information system. Genetic information is an obvious example: the spatial ordering of the base sequence is not random; it is the exact ordering that provides information for the transmission of genetic characteristics in reproduction. Yet the concept of information applies also to other areas such as mental processes, intelligence, social structure, political organization, law, literature, science, art, music, etc. Indeed, the concepts that are characteristically associated with life such as organization, order, direction, purpose, motivation, will, and so on, can be reduced to the basic concept of information. Our contention is that any element in life that is not matter and energy can be reduced to information. The separation of the three elements is not fundamental; it is merely for the convenience of discussion.

In this view the basic laws of life are those of the three basic elements. The laws governing matter and energy except in high energy phenomena are completely given in physical science. These laws are obeyed by matter and energy in living systems and no other laws of matter and energy in living systems (such as the vital force) have been found. This part of the physical basis of life is well known and generally accepted in modern times. It is the basic laws on information that are at the focus of our attention.

Matter and energy are conserved but can be transformed. Information can be transformed but is not conserved. Thus besides the law of transformation there appear the laws of creation and dissipation of information. Our view is that these basic laws are already present in physical science and therefore physical science provides a complete foundation for establishing a system of theoretical biology. On the other hand from these basic laws we can logically derive a series of secondary principles that are not familiar in physical science but are capable of explaining the characteristic features of life. A large part of our efforts is thus involved in the derivation of the secondary principles and the use of them to explain the phenomena of life.

The basic law of <u>creation of information</u> may be found in the theory of fluctuation in statistical physics. Any spatial or temporal arrangement or relation of entities may be realized by random fluctuation. The theory enables us to calculate the probability of occurrence of such events.

The basic law of <u>dissipation of information</u> may be found in the second law of thermodynamics which asserts that any organized system tends to become disorganized, thus reverting to the natural random state.

Concerning the abiogenetical origin of primitive organisms one used to think that the constituent parts of an organism were present in the "primordial soup" and they came together by fluctuation to form an organism. It is true that such formation is permitted qualitatively by the law of creation of information but it is equally certain that such formation is prohibited quantitatively by the law. The probability of forming a highly organized system by high order fluctuations is extremely small, too small to be of any significance. Even if such a formation is realized it will not last long and will disintegrate immediately according to the law of information dissipation. Thus mere fluctuation cannot create life and in the physical world in general there is no "sign of life". However, by a special mechanism, the above situation may be expected and an organized system with an elaborate information content may come into being, may perpetuate itself and may develop to higher complexity. When such a situation arises we see the signs of life. When many signs of life have been developed, the totality of them is what we call life phenomenologically. Thus life is a historical product, not a philosophical entity. The nature of life does not exist before the origination of life; it is the by-product of the origination and evolution of life and can be discussed only in the context of the latter.

2. Origin of Life

In Paper I we discussed a very special set of catalytic synthetic chemical reactions, in an "infinite nutrient medium", in which the number of a set of specified molecules increases in time exponentially. Such a system we call an exponentially multiplying system. The in vitro replication of DNA may be regarded as an example of it. This phenomenon of exponential multiplication is exclusively physical and can be understood completely by the existing physical laws. In particular, chemical reactions obey the second law of thermodynamics and the multiplication is a manifestation of the law.

Yet a new point of interest appears here. The information associated with the structure of the specified molecules is now propagated with the multiplication of the molecules. Ordinarily, according to the law of information dissipation, a highly complex molecule tends to breakdown and the number of it tends to decrease. In the exponentially multiplying system there is an opposing trend to increase the number. When the rate of multiplication exceeds the rate of disintegration, a steady state of the <u>existence</u> of an information system is established for the first time in the physical world. Although the multiplying molecule is hardly life, its appearance may be regarded as the first sign of life. In Paper I we considered exponential multiplication as the first and most basic characteristic of life; all living systems reproduce and multiply exponentially.

Once an exponentially multiplying system comes into being we may imagine the next step of development as follows which may take place according to the basic laws. The copying process in the multiplication of the molecules may incur errors by the law of information dissipation; the reproduced copies may not be exact replicas and some may well be waste products. As the "mutation" proceeds a large variety of waste products may be produced. Once the number of the waste product species becomes very large there is a possibility by random fluctuation that some species may happen to be useful in another way, say, to be attached to the multiplying molecule serving as a protective coat. When such a useful waste product becomes attached to the multiplying molecule, the latter, better protected from information dissipation, can multiply faster. After a number of generations, its population overwhelms other species and the steady state of the existence of the information system is now dominated by this more complex species. This system is closer to the living systems we now know. We may identify the multiplying molecule with the genetic system, the useful waste product with the somatic system, the combination with the organism, the multiplication process with the replication of the genetic molecule as well as the reproduction of the organism, and the garbled multiplication process that leads to the production of the useful waste product with the transcription of the genetic molecule as well as the growth of the organism (3). Of course, modern living systems are further complicated by the addition of a translation process which does not affect the essence of our argument. It is quite possible present living systems originated in this way. The system evolved at this stage has a concomitant linearly multiplying system, the second characteristic of life mentioned in Paper I, in addition to an exponentially multiplying system.

Once such a system has evolved it may change by random fluctuation. Those changes that make the system multiply more efficiently will dominate the population of the species after a number of generations and will be incorporated into the steady state of the information system, this being the well-known principle of evolution. Thus more complex systems may develop, embodying the other characteristics of life discussed in Paper I, i.e., self control system, self-maintained system, motor system, sensory system and nerve system. More complex living species may be developed this way.

This view on the origin of life differs from the conventional views by asserting that the genetic molecule appears before the organism. The conventional views such as that surrounding the concept of the microsphere, place the emphasis on the origination of the individual organism, largely the somatic part of it, from inorganic material. We have already mentioned the difficulty of the small probability of such a high order fluctuation. A more serious problem is how to make the "primitive organism" reproduce itself, for without reproduction the "organism" cannot perpetuate itself. Since the reproduction mechanism of a viable organism, no matter how primitive, is very complicated, it cannot originate all at once by random fluctuation. Nor can it appear gradually for the function of reproduction must work as a whole; any component of it by itself has no "survival value". These difficulties are eliminated in our approach. The origin of the difficulty of the conventional view is to think of life as an a priori entity and of reproduction as one of its attributes. The nature and origin of life thus become inextricably difficult to understand. The difficulty disappears when we regard life as an a posteriori manifestation of the multiplying process which is physical and needs no explanation.

If life is regarded as an a priori entity the origination of the genetic information such as that contained in the ordering of the base sequence becomes a problem. Not surprisingly the quickest way to answer the question phenomenologically is to resort to the theory of Divine Creation. In our view, genetic information does not exist a priori. The self multiplying molecule may well be a nonsense molecule carrying no useful information (most of presently known genetic information is useful information). The nonsense molecule suddenly becomes "sensible" and whatever information it carries suddenly becomes useful the moment the waste product turns out to be useful and becomes associated with the multiplying molecule. Thus genetic information appears a posteriori and is the result of a historical process, understandable by an evolutionary analysis.

In the phenomena of life other information systems, besides the genetic constitution of the organism, such as the mind, intelligence, political organization, social structure, law, ethics and so on, all appear to have an information content (a genome) that seems to have an a priori existence. The origin of these genomes thus becomes a problem and historically the theory of Divine Creation is the most often answer. Thus the mind and intelligence come from a Divinely Created soul, the king's power derives from God, the laws reflect the Divine will, and so on. In our view information does not exist a priori, it appears by random fluctuation and it is incorporated into the genome by evolutionary processes. All these genomes, like the genetic genome, are subject to evolutionary analysis in manners such as that discussed above. By the historical study of the origin and evolution of these information systems we also learn at the same time the very nature of these systems. And this is the only way the nature can be learned. Some of the problems along this line will be discussed later.

3. Principles of Information Safe-keeping

Since information is a basic element of life and it tends to dissipate naturally, its safe-keeping is an important concern. Any mechanism helping safe-keeping of information will improve the efficiency of multiplication and thus will be incorporated into living systems. Many properties of life may be understood from this aspect of the evolution process and a number of principles for various mechanisms of information safe-keeping may be formulated.

1. Principle of _isolation_. Examples: The appearance of membranes and walls, the cell structure of higher organisms.

2. Principle of _redundancy_. Examples: double-stranded DNA, diploid chromosomes and two sexes. "Why sex?" and "why two sexes?" are problems subject to scientific analysis. The answer: the redundancy of sex and sexual reproduction make it possible to remove defective genetic material from some progeny, thus rejuvenating the species. One sex is not enough and three sexes are unnecessary.

3. Principle of _network_. When the redundant components interact with one another, a network is formed. Networks can be parallel or series. In the former a new phenomenon characteristic to life arises, i.e., choice among alternatives (free will?). In the latter another arises, i.e., self control and suppression.

4. Principle of _repair_. Examples: DNA repair in bacteria, tissue and organ repair in higher animals.

4. End of Life

As far as a closed information system is concerned the information content, such as the genetic information of an individual organism, is fixed and can only decrease in the course of time according to the law of information dissipation (somatic mutation by radiation, etc.). Any information safe-keeping devices for an individual organism can only slow the process but cannot stop or reverse it. The consequence is aging and ultimate death. Aging and death are two characteristic features of life. The eventual termination of individual life is a necessary concequence of the second law of thermodynamics. No life can be excepted from this rule unless defective genetic material can be replenished in some way.

From the point of view of the species the ultimate death of individuals is a welcome feature. The physical material the individual is made of is broken down into components and reverted to the natural pool of materials available for the composing of new individuals. The materials circulate continuously in life-death cycles. Each cycle may bring forth new features; thus evolution and progress become possible. If there were no death the materials will be hoarded permanently and there will be no cycling and no progress of the species. The very existence of the elaborate, sophisticated living organism is made possible because of the sacrifices of the predecessors. We now turn to the mechanisms of evolution.

5. Mechanisms of Evolution

Evolution is the process by which changes of the information content of a system become incorporated into the steady state of the system. All changes, according to the law governing the creation of information, come from random fluctuation. Those which increase the efficiency of multiplication (having survival value) become incorporated into the steady state of the system by the shifting of the population distribution; others are eliminated. Still there are a number of different ways changes may occur and we can formulate a number of principles.

1. Principle of substitution. A new, small piece of information, originated by random fluctuation, when useful, substitutes for an old piece in an otherwise unchanged information system. The Darwinian view of evolution in modern interpretation is largely based on the successive accumulation of small, benificial changes.

2. Principle of juxtaposition. Two information systems combine and readjust to form a new information system. This does not occur often in organic evolution of the species, nevertheless, the formation of ecological systems involves the juxtaposition of the components. In molecular evolution before the emerging of the first organism, juxtaposition, such as that between the multiplying molecule and the useful waste product discussed before, is expected to play an important role. In social evolution, juxtaposition seems to be the major mechanism for bringing forth new developments. This mechanism differs from the Darwinian mechanism of substitution in a number of important ways. It makes possible to bring about a drastic, discontinuous change whereas substitution is limited to gradual changes of small steps. Consequently the speed of evolution can be made much faster than otherwise. Moreover, it emphasizes the role of cooperation in evolution whereas the Darwinian view emphasizes competition. It emphasizes the accumulation of information by addition and adjustment instead of by rejection and substitution. Finally it emphasizes that the major problem of survival is a fight against information dissipation not a fight against competing information system. Most of the social maladies of Darwinism can be redressed if it is recognized that besides the Darwinism mechanism of substitution there exists an even more important mechanism of juxtaposition.

3. Principle of differention. One information system differentiates into a number of separate but cooperating information systems. Differentiation of cells, division of labor in a society are examples.

6. Accumulation of Information, Principle of Accretion

The information content of an information system comes into being in the course of time by accretion of newly gained information through substitution, juxtaposition and differentiation. Accretion is a process of linear addition in the chronological order with each addition determined by the immediate contingencies without regard to the past history. Therefore natural information systems do not originate by a grand design and can be expected to contain conflicts, incoherence and inconsistencies among its constituent parts (Original Sin, contradictions). One consequence is that mechanisms of control and suppression become necessary. In the theory of Divine Creation information systems are conceived in a well-organized grand design and are free from such shortcomings. In our view information systems are composed not by authorship but by editorship, and are built not like a well-designed cathedral but like a utility building with one wing piled upon another constructed at different times according to the contingencies of the moment. Not only there is no coherence in outlook but also there may exist conflicts between its parts because the contingencies at different times are completely unrelated.

The linearity of the addition process in the accumulation of genetic information is determined by the use of DNA, which is a linear molecule, as the instrument for storage of information. The linearity of the molecule allows only an information bookkeeping system that is comparable to the daily journal accounting system; the more sophisticated accounting systems such as the ledgers and the financial statements, cannot develop. Once the present life system is committed to DNA it is committed to this primitive accounting system.

The linearity also manifests itself in information retrieval. The only way to retrieve information from a daily journal account is to go through the whole past history day by day. Much of the past information are not needed but it is unavoidable in the daily journal accounting system (it is eliminated in the ledgers). This observation provides us an explanation of the Haeckel law which asserts that ontogeny recapitulates phylogeny, a mysterious manifestation peculiar to living systems. In the development of human embryo, gill, tail, body hair appear and disappear in evolutional order. They are not needed now but the body of genetic information cannot be used to construct a man without having to go through the previous evolutional stages in chronological order because of the linearity of the information storage and retrieval processes.

There is perhaps a similar situation in the development of the brain. The evolutionally earlier brain activities are largely related to instinctive behavior (food and sex) which are superseded, suppressed and controlled by the later brain activities dealing with social behavior. Yet the earlier behavior is not eliminated or modified but controlled or suppressed; it remains in the information storage because of the nature of linear accretion in the accumulation of genetic information. The earlier brain activities become the subconcious which is one peculiar characteristic of man and for which we offer an explanation of its nature and origin.

7. Patterns of Evolution of Information Systems

The outcome of the evolution of an information system due to various mechanisms in response to various contingencies over a long period of time may lead to a present day state of the information system quite different from what it began, so different that it may be beyond recognition. Without elucidating the mechanisms and processes of evolution the present day system becomes incomprehensible and many times Divine Creation may seem to be the only answer. On the other hand there are general patterns of evolution that can be recognized and can be formulated into a number of principles. A knowledge of these principles helps us understand the historical origin of many of the present day information systems. Since one of our main objectives is to explain the origin of the present day information systems in terms of the principles developed here, a discussion of the patterns of evolution helps us to achieve this aim by short cutting much of the formally similar evolutional analysis.

1. Principle of <u>perfection</u>. Information system develops toward the point of maximum efficiency. For example, marine animals develop toward the streamline shapes. The Darwinian principle of survival of the fittest is usually used in this sense.

2. Principle of <u>imperfection</u>. Once maximum efficiency is reached evolution by the principle of perfection stops. Other mechanisms are necessary for continued evolution. Imperfection means

less commitment and makes it easier to change from one form of operation to another and thus to develop new information content. To avoid danger helps survival. Yet many of the human successes are not due to the human ability to avoid danger but due to a human disposition to risk danger. In doing so new discoveries are made and information content is expanded. Conditioned reflex and association are probably due to imperfections of the nerve mechanisms (jumping of wire in the brain). Yet this imperfection made possible a new biological process, i.e., learning which figures importantly in the development of man.

3. Principle of <u>specialization.</u> Specialization improves efficiency and therefore is a general tendency in evolutionary processes.

4. Principle of <u>versatility</u>. Just the opposite to specialization, sometimes versatility is desired. Man is not the best runner nor the best swimmer, but is the most versatile among all animals.

5. Principle of <u>conversion</u>. Purpose is a characteristic of life. It emerges from evolutionary development, usually from purposeless. Once emerged the purpose may change in later evolution. One process of conversion is "from trash to treasure" such as that discussed before in connection with the origin of the transcription process. Another is "from luxury to necessity" exemplified by the use of tools by man. In economic development the conversion of luxury to necessity is the one single internal factor dominating the trend of economic development. The predicted doom of the industrial economy after World War II was based on the quick saturation of the demand on necessities by industrial production. Yet, far from collapsing, the industrial world enjoyed the most prosperous period of development after the War, to the surprise of all economists. The reason is that many luxury items are converted into necessities (automobile, household appliances) and thus the demand for production is greatly increased.

6. Principle of <u>complication</u>. The general tendency of evolution is toward complication even though the complication is not absolutely necessary (much ado for the same thing). One consequence is the increased amount of waste. This is the origin of the current problem of pollution. Any solution of it must take the general evolutional problem into consideration.

7. Principle of <u>inversion</u>. A contributing factor in an evolution process may eventually turn out to become an attribute of the evolved system. One used to ask: why life reproduces (inanimate material does not)? The attribute reproduction of life, in our view, is actually the cause of life. Because of exponential multiplication a new set of phenomena comes into being which we call life. Without reproduction there would be no life in the first place. This pattern of inversion of cause and "effect" occurs frequently in evolution and an understanding of it helps answer some of the more puzzling questions.

One used to ask: why should civilized society suppress sex which is biologically vital and morally neutral? The answer is that without the suppression of sex civilized society will not appear in the first place. Civilized society is not an a priori existent, or a Divinely Created, information system no matter how much one wishes it to be. It comes into being by a series of evolutional processes in which the introduction of sex taboo played an important part. It appeared by random fluctuation but because of it a large part of human energy was diverted from sex to other activities which eventually developed into culture and civilization. Now we are dependent on civilization and thus on the factors that brought it into being.

One used to ask: why should rational man have anything to do with irrational religion? The answer is that without the latter the former will not be here in the first place. Rational man is not an a priori existent information system and is a product of the civilized society. Irrational religion acted like an enzyme in bridging over potential barriers in the process of formation of human society. In particular the ecumenical religions are instrumental in bringing about the transition from tribal societies to universal human society from which the rational man emerges.

Geographisches Institut
der Universität Kiel
Neue Universität

On the other hand the conflict of rational man and irrational religion is real but it is not unexpected (Sec. 6) and can be resolved by evolutional processes just as in all other information systems. It is not a philosophical problem but a historical problem. This discussion brings out the contrast between history and philosophy. Life, man and society are essentially historical problems. The theories of Divine Creation substitutes philosophy for history. Much of the current intellectual confusion has an origin of a similar nature.

8. Principles of Revolution

Evolution sometimes results in an apparently drastic, discontinuous change of the information system. When this happens it is called a revolution. Revolution is not frequent in organic evolution of the species. On the other hand in molecular evolution before it and social evolution after, revolution may play more important roles. A number of principles on such changes may be formulated.

1. Principle of redundancy. The mechanism of gene duplication in evolution is a suitable example. A gene is duplicated; one of the two serves the original function and the other can afford to be changed in a variety of ways in succession until it becomes a drastically different new gene serving a new function.

2. Principle of juxtaposition. By the very nature of juxtaposition, a drastically new information system is formed by the process. Social and cultural evolution largely involve juxtaposition. The speed of information changes and accumulation is much faster than in organic evolution.

3. Principle of subversion. The new information system establishes itself by making use of the machinery of the old information system to serve its own purpose. The way a phage attacks a bacterium, converting the bacterial material into phage progeny, is a good example. Political subversion follows the same pattern.

4. Principle of transfiguration. Not all revolutions are subversive; some are transfiguring. The necessary elements of the new information systems were first developed within the old information system as a natural outgrowth of it. Once the necessary elements of the new system are so developed, even for a different purpose, it is not difficult for them to be reorganized into a new system.

In the evolution of man the change from quadruped to biped together with the associated change to the upright posture and the change of the function of the fore limbs to tool handling are revolutional changes, drastic and discontinuous. Very likely the changes are prepared in a period when the precursor of man was arboreal. All four limbs were still used mainly for locomotion but the fore limbs gradually developed the skill of grabbing tree branches for the same purpose. At the same time the upright posture gradually developed in arboreal life. These changes can be seen in the present day arboreal primates. Then the forest vanished and the precursor of man was forced to live on ground and was forced to learn to walk on the hind limbs to keep the upright posture, which he quickly succeeded. The ability of the fore limbs to grab objects then developed into the skill of tool handling. Thus the necessary elements of the new information system were first developed in the old system for the old purpose of arboreal locomotion. Once given the suitable opportunity they transfigured to form a new system.

The development of the theory of relativity, the quantum theory and even the experimental science at Galileo's time are revolutional changes. How they came about are interesting questions in history of science. We can see the workings of the principle of transfiguration in all of them even though we do not have space to elaborate here. The change from the classical tonal music to modern atonal music (Schönberg) is a revolutional change. Again we can see the change being first prepared in the classical system in the compositions of Wagner who worked within the framework of tonal music but experimented with changes of keys in rapid succession in such a way that tonality is practically lost. Thus elements of atonal music were developed first in the framework of tonal music.

9. Applications

Our purpose of developing the theory sketched above is not only to formulate a theoretical system, grounded on the well established knowledge of present day physical science, that will enable us to understand the phenomena of life, but also to use the theory in a similar way to elucidate the origination and evolution, and therefore to understand the nature of many other information systems associated with life such as the man, the mind, the society and the cultural developments which used to be understandable largely by the theory of Divine Creation. Our aim is to analyse the history of accretion of information, originated by random fluctuation, selected by various mechanisms, developed through various patterns of evolution and revolution according to the principles discussed above. Some of such studies will be described in a more detailed paper (1).

10. Epilogue

The nature, origin and evolution of life, man, mind, society and culture are basic problems in intellectual pursuit not only for pure theoretical interest but also for establishing a foundation to deal with practical problems. Since time immemorial man has engaged in contemplation on these problems. Not surprisingly the basic concept information we introduced here had its counterparts in previous works going back to ancient times. ANAXAGORAS in fifth century B.C., departing from the materialistic thought of one Prime Substance, proposed one Mind as the unifying principle which in the beginning turned chaos into order. The concept had its origin in HERACLITUS' logos, an immanent reason at work in the cosmic processes. It was developed later in various stages and dominated Western thought in philosophy through the Greek thinkers and in religion through its appearance in the New Testament (the Word) and its influence on the Trinity Doctrine. In the Eastern world the concept of an overall organizing principle (the Tao) transcending the material (the Chi) originated in ancient thoughts including Confucianism and Taoism. It became the central idea in Neo-Confucianism (Taology), from the eleventh century on, in an attempt to unify all knowledge from cosmology to ethics.

With our study in the background we can now identify the ancient concepts of logos, the Word and the Tao with the scientific concept of information which goes beyond matter and energy and organizes the latter into life and life-like information systems. With this identification we also find unity in science, humanities and religion. Thus we may close by quoting John 1 (1-5) with fresh insight.

References

1. FONG, P.: DNA, RNA and the Physical Basis of Life, a collection of papers to be published by Gordon and Breach, New York: Science Publishers.
2. FONG, P.: J. Theoret. Biol. 21, 133 (1968).
3. FONG, P.: Proc. Nat. Acad. Sci. (Washington) 58, 501 (1967).

DISCUSSION

BREMERMANN:
Manfred EIGEN's new theory of chemical evolution parallels the idea of section 2 and puts them in a quantitative, mathematical and experimentally testable form.

FONG:
Thank you. I should like to study this theory when published.

STEIN (Communicated by SIMON): I would try to make an objection against certain ideas concerning the accumulation of information. The recording is regarded as a linear process, each addition being independent of the previous ones. A necessary "recapitulation" of both useful and useless information during ontogenesis is explained by the linearity of information retrieval. Although DNA is a linear type recorder, information is impressed simultaneously all over it and unlikely to a tape recorder, it can be impressed and regained at several sites at the same time. DNA information is available not only by chronological-type linear lecture, but also by ledger-type simultaneous retrieval on the molecule. Evolution (phylogeny) cannot be regarded as successive inscriptions on an unimpressed part of the DNA, but rather as alterations of the information contained by this molecular recording. Thus the informative evolution would rather go from non-sense molecules to useless information, which changes to useful and more useful information by mutations. I only reluctantly can agree to denominate "the survival of the fittest" as a principle of perfection, as the admittance of perfection in biological efficiency implicitly brings us to admit purpose or at least an "a priori" schedule, more or less followed by an organism. The same objection can be made to the opposite principle, that of imperfection. - This is in contrast with the authors careful attention to distinguish "a priori" philosophical entities in classical biology (e.g. "life") from his view, regarding both life and genetical information as an "a posteriori" phenomenon of historical evolution.

FONG:
I am glad that you point out that DNA information can be used in a way like a ledger account is used. This happens when the cell carries out the daily housekeeping functions such as the secretion of hormones, etc. When I say in the paper that DNA information is used like a journal account I refer to the process of embryological development. - It is true that evolution may involve alterations of existing information. Yet addition of information is another possible mechanism of evolution. - Ontogeny does not necessarily recapitulate everything in phylogeny just as some of the bricks and walls of a lower floor can be removed without tumbling the upper floors. On the other hand some elements cannot be removed and will have to be recapitulated. - The terms perfection and imperfection are, of course, used in an "a posteriori" sense. For example, perfection in water means imperfection on land; there is no a priori sense of perfection.

BRUNNGRABER (see page 20):

FONG:
It is very much desired that information would flow from protein to DNA (against the Central Dogma) so that a useful protein could be codified readily on DNA and incorporated into the genome all in one stroke (instead of by trial and error variations). However, in the absence of evidence for this process to occur, we have to depend on the less desirable yet definitely workable mechanisms to explain the origination of life.

YOCKEY:
Your definition of information is vague. You should put this in mathematical form and certainly other workers such as WIENER, SHANNON, and QUASTLER deserve mention in the references. A sodium chloride crystal is an orderly structure which is created out of a disorderly substrate, namely a saturated solution. It grows and reproduces itself. However, as I pointed out, it con-

tains very little information and it is not alive. The concept of information does not apply to mental processes, intelligence, etc. The importance of the "basic law of creation of information" and the "basic law of dissipation of information" was discussed in my paper. It would be most valuable if you would contribute some specific analysis on this subject. An exponentially multiplying molecule is a sign of an autocatalytic reaction – of which many are known in chemistry – and is not by itself a "sign of life".

FONG:
I defined my concept of information clearly in the beginning of my paper (see reply to ANDREW). It was so defined that it may be applied to more complex systems such as mental processes, intelligence, etc., beyond those well known to the physical scientists and communication engineers. For this purpose it is deliberately left qualitative without any attempt at quantification. A definition is meaningful and useful only in connection with the theory that follows. For the purpose of this paper no other definition of information is needed. It is obviously not the information of WIENER, SHANNON and QUASTLER and is indeed used for a completely different purpose (see reply to ROSEN). Since the two are different scientific entities, to avoid unnecessary confusion perhaps I should use a different name such as $\lambda o \gamma o s$ or Tao. But among the English words information is the most appropriate. "Random" may be defined by the physically most probable state or the state of maximum entropy, any deviation of which is considered to represent information. In other words, random is a state determined exclusively by the laws of matter and energy (dynamical and statistical laws). (Any deviation from it being information we consider the three basic elements of life being matter, energy and information.) In most cases of practical significance this definition of random is not much different from that of Webster's Dictionary; yet it will also include cases like the sodium chloride crystal which has no practical significance. The question whether a growing crystal or an experimentally multiplying molecule should be considered as life or not is discussed in reference (2) of my paper. – The importance of the basic laws of creation and dissipation of information lies in the establishment of the vital link between physical science and biological science which I have done in my paper.

MEL:
MOROWITZ has discussed spontaneous development of complex function and structure in initially disordered systems (H.S. MOROWITZ, "Energy Flow in Biology", Academic Press, N.Y. 1968). He argues convincingly that, far from being highly improbable, such developments are natural consequences of (energy) flows through bounded open systems containing appropriate elements.

FONG:
MOROWITZ'argument may be applicable to the origination of a complex structure (such as the microsphere), but any mechanism leading to the origination of a structure is not likely to lead to the origination at the same time of a self reproducing mechanism for that structure. The two are distinctively different; any theory explaining one is not likely to explain the other.

ROSEN:
I think that everyone interested in organization must decide for himself whether he believes the view, inherited from physics (and indeed from the physics of very simple systems), that the terms "random", "homogeneous" and "maximal entropy" are synonymous. Biology is full of systems whose autonomous dynamics drives them towards a particular end-state which is inhomogeneous. On the basis of ordinary physical ideas, this (asymptotically stable) end-state is by definition the state of "maximal disorder" or "maximal entropy" for the system in question, but it is not homogeneous, and a physicist would typically call it non-random or ordered (or in the terminology of this paper, to say that it contains "information"). The Turing systems for differentiation and pattern generation are perhaps the simplest examples of this. My personal feeling is that, to be meaningful beyond the simple and atypical systems with which physicists have dealt, entropy must be considered not merely as a function of state (i.e. independent of any dynamical laws) but must take explicit account of the dynamical laws which govern the behavior of the system.

Without this, I believe that arguments based on "entropy" or "information" in complex systems tend to be misleading.

FONG:
Information and entropy are terms often misused. In this paper the term information is defined precisely in a particular way and is used only in the defined sense. No analogy to any simple physical system is ever used. The concept is introduced precisely for the analysis of complex systems by rigorous logical procedures. There will be no misunderstanding if it is kept in mind that the concept information is used only for the investigation of its origination, preservation, accumulation, utilization, dissipation and evolution. All these discussions can be made dissociated from the confused concept of entropy and in this paper little attempt is made to exploit the concept of entropy. ROSEN's paradox can easily be explained as follows: The biosphere is not a closed system and is not in an equilibrium state; therefore the entropy is not maximized. It is rather in a steady state of energy flow in the MOROWITZ sense.

ANDREW:
The statement that "The basic law of creation of information may be found in the theory of fluctuation in statistical physics" is liable to be misleading. Information, in SHANNON's sense, is only strictly defined in terms of a set of symbols agreed between the originator and recipient. It is reasonable to stretch these definitions a little in order to treat such things as nervous and humoral communication, and the transmission of genetic data, where it is clear from observation that some sort of agreement over the meaning of the "symbols" has been reached by the parts of the systems originating and receiving them. Fluctuations in physical systems do not create information in this sense. The main conclusions of the paper are not affected by the somewhat loose treatment of "information", but it is important to be rigorous since some other theorists have gone badly astray over this sort of thing. — The view that a genetic molecule capable of replication must have preceded other manifestations of life seems very reasonable in view of the arguments of the paper. The suggestion that genetic information is organized in a sub-optimal way because of the linearity of the DNA molecule is very intriguing. A species which suceeded in organizing the information more efficiently could speed up its rate of evolution markedly. I would be interested to know whether FONG thinks this is a possibility, either as a result of "natural" evolution or of human "genetic engineering".

FONG:
The term information used here is defined in the beginning of the paper as any spatial, temporal arrangement or relation of entities that is other than random. Therefore it is not the information in the SHANNON sense and any Shannonism should be forgotten here. Since fluctuation may lead to a particular spatial, temporal arrangement other than random, we say information may be created by fluctuation. This definition of information is very general. In spite of the lack of specificity the definition is nevertheless sufficient to organize a large body of biological principles. A more narrowly defined concept, of course, is expected to lead to more specific conclusions. — The DNA-RNA genetic system is so comitted to what it is that it is unlikely any breakthrough in its improvement may happen either by natural evolution or by genetical engineering (just as a horse is too committed to be a horse that it is unlikely to evolve into a whale). Yet breakthrough in better information organization may happen beyond the DNA-RNA system. Beadle once gave an example of transmission of living characteristics by a "mechanical" means independent of the DNA-RNA system. The other mechanisms of information preservation and transmission are subject to drastic changes with catastrophic results. One example is the explosive evolution of man overwhelming all other species of life because of the use of language and printing for information transmission and preservation.

LOCKER:
Your theory of life, especially by enumerating several primary and secundary principles, which are certainly more than purely phenomenological and present a hint at underlying mechanisms, is

very plausible. Nonetheless, you offer some ideas which deserve discussion. If any spatio-temporal organization that differs from a random arrangement should be considered as "information", several conditions must be fulfilled: It must something else exist for which the differing (or deviating) configuration has meaning, i.e. information. It is hard to understand that this organization has arisen randomly only if, as you suppose, the random event is the criterion for the distinction of information from non-information. Although, in stating that the genetic molecule cannot exist before the organism, you correctly touch upon a basic paradox - which, for instance, has also been viewed by PATTEE in his paper - you nevertheless believe to eliminate its logic and ontologic difficulties by referring to a historical process as opposing an a priori datum. In that mere fluctuations cannot create life, as you correctly say, the lacking additional mechanism is just the functioning of the system which must preexist before information can arise (or even, in connexion with which information only can originate). If your sentence, that the nature of life (i.e. simply, life) does not exist before the origin of life, should not be a pure tautology, it seems to be intended by this phrase to circumscribe or circumvent the difficulty related with a double nature of life, namely to occur within time (so-called objective, historical time) and to simultaneously create time during its emergence (subjective or operative time). Of course, your hypothetical description of the origin of life is quite fascinating and certainly contains high amount of probability, although the permanently underlying assumption of random fluctuations as quasi motive processes of evolution is probably only valid for small steps (i.e. microevolution), but not for the global evolutionary unfolding of the biosphere. Also the supposition that a nonsense molecule becomes sensible can only be understood under the guidance of a processuality which is so universal as to overcome intrinsic contradictions or conflicts. Thus, although you have found a beautiful expression, the question is whether there is no more contrast between "autorship" and "editorship" than between two principally distinct but also intrinsically well-matched points of view. You still approach another paradox in your paper by underlining the fact that the beginning of information is now beyond recognition. This, I would like to add, is principally true so long as the observer (or theoretician) poses himself within the process described and does not try to assume an outside (or above) position. Such a position can, under certain circumstances, be achieved only conceptually, but can probably be modeled by ingenious simulation experiments. Your epilogue shows, however, that you consider information as a qualitative, yet metaphysical, entity and in that respect you probably come near to the root of the problem.

FONG:
Let me explain my view by an example. In an in vitro DNA synthesis system designated by A any DNA synthesized is said to contain information because it has a definite base sequence which is not random. If the template DNA is removed (this is a different system designated by B) and polymerization is allowed to proceed the DNA synthesized has a random sequence and is said to contain no information. The information of the DNA in system A is meaningful when compared with that in system B. It is true in system A the information originated from the template, not from a fluctuation process. Phenomenologically in many cases we can trace the source of information to some other entity. But the question of the origin of the information of that other entity (e.g. the template DNA) now arises and will have to be answered. My view is that the ultimate source of information is fluctuation. For example, the base sequence of the template DNA results from many steps of mutation, each of which is related to a random process of fluctuation. The question of the ultimate source of information is not trivial. In fact it is the basic and central philosophical and theoretical problem. The essence of the theory of Divine Creation is that the ultimate source of information has a separate, independent existence beyond and before the material system, this being the main point of the Johannine Prologue. In fact this theory is just about the only theory generally known dealing with this problem (in Taology the Tao is assumed to have a transcendental existence above the Chi and thus the theory is not different from that of Divine Creation) and as far as the social problems are concerned this theory is still unconsciously the keystone of the mental framework for the majority of intellectuals despite their disclaimer on Divine Creation. My main purpose is to understand the origination of infor-

mation (biological and social) without having to assume an a priori transcendental existence. The set of secondary principles are developed to explain phenomenology in terms of a set of basic concepts and the set of primary principles are formulated to trace the origin of the basic concepts to the body of established principles of physical science. The theory is thus a closed and complete system. When completely developed, what used to be considered as an a priori datum can generally be explained by an historical process without logical and ontological difficulties. Section 2 illustrates the point that the "functioning of the system" which is the origin of the ontological difficulty need not be preexist but can be evolved out of "nonsense". Although random fluctuation is commonly associated with small-step evolution, the principle of juxtaposition implies that it may also lead to catastrophic evolution. My statement that the beginning of information is now beyond recognition is made in an historical sense as a matter of fact (such as the burning of Persepolis) not in a philosophical sense (such as the sinking of Atlantis) and thus involves no paradox. It is true that the concept of information is treated here on a qualitative basis but this does not imply the investigation is preliminary and unfinished. Unlike the solution of a differential equation the investigation here resembles more the study of topology which is final and absolute.

Thermodynamic Potentials and Evolution towards the Stationary State in Open Systems of Far-from-Equilibrium Chemical Reactions: The Affinity Squared Minimum Function

H. C. Mel and D. A. Ewald

Abstract +
The "evolutionary" characteristics of several quadratic functions (based both on the affinities and on the reaction velocities), and of the entropy production per unit time, have been studied for a number of 2-variable open systems of far-from-equilibrium chemical reactions. The first and second order systems were chosen to include: straight line, loop (network), autocatalytic and disproportionate kinetic features. All of the functions examined are closely related in form to the entropy production though they differ qualitatively and quantitatively in the non-linear domain. By a combination of analytical and computational methods one function, called μ, is seen to have the variational properties of a "thermodynamic potential" for all of the systems, relative to their non-equilibrium stationary states. The "homeostatic-like" stability criterion, $\frac{d\mu}{dt} \leq 0$ is also seen to hold for these systems. The function, a composite property of the system, may be interpreted as a "kinetically-weighted system-free-energy" quantity, which would tend to be minimized during the evolution, in configuration and in space time, of a constrained biochemical system.

DISCUSSION

SIMON (see page 111):
MEL:
We wholeheartedly agree with the interest in comparing and attempting to understand the relationships between the several functions (and their associated variational properties) presented by different authors, including that of KLINE. This is not entirely a trivial exercise, in view of the differences in systems thus far treated, and in points of view of departure - i.e. primarily mathematical, for many, as opposed to more physical-chemical thermodynamic for others (such as ours).

WALTER:
It is probably not be very interesting to examine the behavior of μ for GOODWIN's system, because that system does not correspond to any real chemical system (it can, for example, lead to negative concentrations). What might be interesting however is to examine the behavior of μ for the two-component positive feedback system studied by HIGGINS and the multicomponent positive and negative feedback systems studied by myself (both of these systems describe known biochemical mechanisms and both admit the possibility of limit cycle behavior). It should also be very interesting to examine μ when a system prossessing a single unstable stationary state (that is, a biochemical system oscillating about an unstable focus) switches into a system possessing three stationary states (two stables focuses and one saddle point). I have published results of studies of the dynamical behavior of systems of this type (which also possess known counterparts in biochemical systems).

+ Full paper will presumably appear in BG & Th.

MEL:
We would very much like to see the μ behavior analyzed for such plausible chemical systems, both further to elucidate the properties of the μ function, and for possible insights into (control features of) the processes themselves. Regarding permanent oscillatory systems, at this point it is not clear what behavior the function ideally "should" have. Possibilities for μ include: a) co-oscillations; b) constancy at its minimum (infinite time) value; c) constancy above its minimum value, indicative of a higher level of dissipation, and perhaps with a "rotation" (in concentration space) at this constant value.

ANDREW:
The prospect of applying a unifying principle to the study of a wide range of phenomena is attractive, and the authors show that the Affinity Squared function is potentially extremely valuable. However, I feel uneasy about the use of the word "evolution" in this connection. Is the progression of a system towards a stationary state necessarily to be termed "evolution"? The point of view expressed by ASHBY in his "Design for a Brain" suggests it should. After all, the system in its stationary state is, in a sense, better adapted to its environment, as shown by the fact that it can survive unchanged in that environment. All what we usually mean when we talk about evolution of an animal species is that it becomes better able to survive in its environment. – Despite these arguments, it is impossible to avoid a feeling that evolution should be more "clever", or goal-directed in some "higher" sense. If a house collapses into a heap of rubble, it has reached a state which is more stable than its previous one, yet it would hardly be said to have "evolved". If it is, as houses usually are, a part of a larger system involving people, its collapsed state may not really be a stable one, although it appears to be when only the sub-system comprising the house is considered. In the context of the wider system there is justification for the intuitive feeling that the collapse does not constitute evolution. – In discussing living systems, or systems having constituent parts which are alive, the idea of "evolution" is a complex one. It would be interesting to know whether the Affinity Squared function is, in principle, applicable to systems in which apparently-stable subsystems are liable to be bulldozed aside to make way for apparently-less-stable ones.

MEL:
We are of course using the word "evolution" in a more restricted sense than the Darwinian one (as suggested by our frequent use of quotation marks). However reaction systems do evolve from initial to final conditions, and there is a good thermodynamic precedent for such usage. (Note, e.g. P. GLANSDORFF and I. PRIGOGINE, Thermodynamic Theory of Structure Stability and Fluctuations, Wiley-Interscience, New York, 1971, Chapter IX: "The General Evolution Criterion".) For closed reversible systems the original thermodynamic evolutionary principle – the Second Law – does indeed describe movement from order to "rubble", and as your comments imply this is of no interest as a life-organizing principle. For "evolution" in constrained irreversible systems, to non-equilibrium steady states, minimization of the affinity squared function should be quite compatible with a movement away from "rubble" (decreasing entropy of the system) in the same sense that the "minimum entropy production principle", in its linear domain of validity, is compatible with and characteristic of this more interesting kind of evolution. (I. Prigogine, Introduction of Thermodynamics of Irreversible Processes, Third Edition, Interscience-Wiley, New York, p. 76.) – Even in a more Darwinian sense, however, thermodynamic potentials have long been considered as possible "evolutionary indicators". (note KATCHALSKY and CURRAN's remarks +, on the past belief of many theoreticians that "... the concept of least dissipation is the physical principle underlying the evolution of the phenomena of life.") To investigate such possible behavior for the μ function one could examine, for example, more primitive vs. more evolved forms of a given metabolic pathway (with identical initial and final conditions) for co-systematic changes in the μ function, e.g. in values for μ stationary.

+ In: Nonequilibrium Thermodynamics in Biophysics, p.234, Cambridge, Mass.: Harvard Univ. Pr. (1963).

Thermodynamic Stability and Spatio-Temporal Structures in Chemical Systems

G. Nicolis

Abstract +

The extension of the local formulation of thermodynamics to include a theory of stability and of fluctuations is reviewed. It is shown that purely dissipative systems such as systems undergoing chemical reactions and transport processes may exhibit instabilities of their steady state solutions provided they are maintained far from equilibrium. The subsequent evolution to a dissipative structure is studied for a simple model system. The biological implications of the results are discussed with relation to 1. the possibility of existence of localized dissipative structures as a stabilizing mechanism and 2. concentration waves as a plausible mechanism for propagation and transmission of information in the form of chemical signals   being of great importance during the process of growth and development of higher organisms.

Reference

HERSCHKOWITZ-KAUFMAN, M., NICOLIS, G.: J. Chem. Phys. (1970).

DISCUSSION

ROSEN:

It seems to me that the attempt to maintain the formalism of thermodynamics when dealing with systems far from "thermodynamic equilibrium" generates a lot of unnecessary trouble. As is quite clear from the paper under discussion, and many related works in "non-equilibrium" thermodynamics, the equilibrium methods simply do not generalize, for clear-cut and specific mathematical reasons. The problem concerned are <u>dynamical</u> problems not thermodynamic ones, and the methods appropriate for dealing with them most easily are those of stability theory (also called "qualitative theory of differential equations"). It is, of course, important to relate the results of stability analyses to the kinds of thermodynamic argumentation which work for "near-equilibrium" systems, but this relation, in my view, will have to come from the direction of the more general, and not from the more special theory. My remarks on "entropy" in connection with FONG's paper are perhaps appropriate here also.

NICOLIS:

It seems to me that most of the arguments presented by ROSEN are in fact in favor of the thermodynamic approach. It is true that stability far from thermodynamic equilibrium depends on a very complicated interplay between purely thermodynamic and kinetic quantities and that it does not seem possible to build a theory of stability independently of kinetic considerations. We are in the presence of a striking difference with the universal validity of the laws of classical thermodynamics to which one is accustomed in the neighborhood of the equilibrium state. – Clearly, this fundamental difference between linear and far from equilibrium domains of irreversible processes could not be sorted out by means of the qualitative theory of differential equations alone. In addition, the dependence of stability on kinetics which implies a great multiplicity of possi-

---

+ Full paper presumably appears in BG & Th.

bilities for change and evolution, is precisely the type of situation which should be of special interest in fluid dynamics and biology. An energy or mass flow may result in the emergence of a spatial, temporal or functional order corresponding to a decrease in entropy, an increase in entropy production or else to some other functional advantage as compared to the original unstable steady state. A striking example of such functional advantages has been developed very recently by EIGEN in his theory of evolution of biological macromolecules (M. EIGEN, Naturwiss. 1971). The main point is that far from equilibrium all these new possibilities are permitted and are in fact exploited by the system. On the other hand close to equilibrium the same system obeying the same differential equations would follow inevitably the theorem of minimum entropy production and would tend to _destroy_ any ordered structure initially imposed on it. In conclusion one may say that thermodynamics by the generality of its statements, its power of unification and its ability to describe complicated situations in terms of a few simple concepts, remains an outstanding tool of investigation. The fact mentioned by ROSEN that the equilibrium methods simply do not generalize in nonlinear situations, far from being a flaw of the theory described in this paper, illustrates one of its most interesting features. It is not claimed that the local thermodynamic theory is the only approach to non equilibrium situations, although I do not exactly know the type of more general theory ROSEN seems to imply in his comment. In either case I should like to emphasize most decidedly that local thermodynamics is _not_ a trivial or straightforward extrapolation of equilibrium theory. New concepts, e.g. the explicit incorporation of fluctuations in the macroscopic description have been needed in order to account for the essentially statistical character of the evolution in the neighborhood of instabilities.

PAVLIDIS:
The derivation of concentration waves on the basis of thermodynamic considerations is a very interesting and important piece of work. I would like to ask about the possibility of different types of waves in the same medium depending on the initial and boundary conditions and in particular about substantial changes in their frequencies. Such behavior would be of great interest for those who study biological rhythms.

NICOLIS:
This is a very interesting question which is presently under investigation. We have not yet analyzed the influence of initial and boundary conditions on the various properties (such a periods) of the wave. However in the case of localized steady state structures preliminary studies show a lack of uniqueness of solutions of the conservation equations for fixed boundary conditions. The change in the initial conditions seems to influence the number of peaks and the localization of the patterns.

Optimal Adaptation of the Metabolic Processes in the Cell

D. Detchev and S. Teodorova

Abstract

In a thermodynamical description the cell can be treated as a highly non-equilibrium optimal self-controlling system. The natural functional according to which the cell optimizes itself is formulated in connection with the principle of self-maintenance of living structures. The maximum principle of PONTRJAGIN is used to determine the behavior as a non-classical variational problem. In order to describe the cell not only as a self-regulating but also as a self-adapting system a set of additional equations for the control parameters has to be derived. From the basic theorem of the maximum principle and from its connection with BELLMAN's method of dynamic programming the conclusion can be drawn that the dissipation of free energy or the rate of entropy production are functions of the state and of the controlling parameters as well. The flow of energy used is maximized along the optimal trajectory. Describing the thermodynamical constants as a function also of the control points offers the possibility for non-statistical thermodynamic consideration.

Rererence

DETCHEV,G., MOSKONA, A.: In: LOCKER (Ed.): "Quantitative Biology of Metabolism", p. 195, Berlin, Heidelberg, New York: Springer 1968.

DISCUSSION

SIMON:
In simultaneously commenting on the contributions made by DETCHEV and TEODOROVA; MEL and EWALD; GROSS and KIM, it seems to me that these are closely related to the work by GOODWIN on the theory of cellular control processes. This theory is based upon the existence of a general function G of various concentration variables exhibiting the properties of a Hamiltonian. Is it possible that Hamiltonian-like coupled variables ($\dot{x}_i = -\partial G/\partial y_i$, $\dot{y}_i = \partial G/\partial x_i$) exist also for the function H of DETCHEV and TEODOROVA or for the function μ of MEL and EWALD? What relations do exist between the minimization principles of these two notes; is it possible to obtain one minimization principle from the other or do they exclude each other? These problems are complicated by the fact that the three approaches mentioned refer to somewhat different types of variables. DETCHEV and TEODOROVA (in their extensive paper read by me) treat the mole numbers of substances as separated from the control parameters; they define the function $p_1(t), \ldots, p_n(t)$ (components of an n-dimensional vector) which with $y_1(t), \ldots, y_n(t)$ (n-dimensional phase space of metabolites) give pairs of coupled variables (y_i, p_i) as required by GOODWIN. On the other hand the chemical affinities A_i appear both in the function H of DETCHEV and TEODOROVA and in μ of MEL and EWALD. The most important test of the validity of the theories would be brought about by applying the different minimization principles to concrete instances, like the autocatalytic diffusion-coupled systems investigated by GROSS and KIM. It is perhaps difficult to find out in these models the correspondences to the control parameters, but a μ-function seems readily computable if for the A_i-

substances the affinities or chemical potentials can be given.

GROSS (see page 70):

MEL (see page 107):

DETCHEV:
Our work cannot be compared with GOODWIN's, since this author describes the metabolic system in a statistical sense by averaging the trajectories of the elementary oscillators and by even reducing the control effects to a purely statistical background whereas in our studies the cell is viewed as a self regulating system by aiming at a non-statistical approach in thermodynamics. The equations describing cellular behavior are derived from the theory of optimal control whereby the function H turns out to be analogous to the Hamiltonian H of classical mechanics in a formal way only. H depends strongly on the controlling parameters and minimizes (maximizes) in correspondence with them. GOODWIN's G is an energy analogue of classical mechanics whereas our H denotes dissipation of free energy. As to the connection formally existing between $\partial H/\partial u_\varrho = 0$ in our work and MEL's $\partial \mu/\partial t \leq 0$ one has to point out that MEL's stability criterion has nothing to do with processes of self-control since it has been elaborated in connection with separated unrelated chemical reactions. The coupled variables (y_i, p_i) which determine our system in an absolute way have quite another meaning than the similarly coupled variables in GOODWIN's papers. In our work y_i denotes the mole numbers of the various substances that take part in metabolism and p_i express the chemical potentials of these substances whereas in GOODWIN by means of x_i and y_i, respectively, are expressed the quantities of mRNA and synthesized protein.

On the Generation of Metabolic Novelties in Evolution

R. Rosen

Abstract
In this paper, a number of dynamical evolutionary models are considered from the structural and functional point of view. It is shown that these models differ in character from other kinds of dynamical models of biological processes, and seem intrinsically unable to account for the generation of evolutionary novelties. A few suggestions are offered for widening the scope of these models.

I believe it is a fair statement that our understanding of the dynamics of evolutionary processes, when viewed in terms of phenotypes, has lagged far behind our understanding of other kinds of biodynamical processes, such as those involved in the study of physiology and of development. At first sight, there does not appear to be any intrinsic reason for such a lag; it was noted very early that there are strong formal similarities between evolutionary and developmental processes (cf. (3), Chapt. 4 for a brief review), while on the other hand, we already have a fairly good understanding of what comprises a dynamical model, or better, a dynamical metaphor, for all kinds of biodynamic and developmental processes (6). Nevertheless, the lag is real, the underlying reason for it profound; the present note is devoted to an exploration of this situation, and the formulation of a few tentative proposals for overcoming the difficulties.

Let us begin by briefly describing the two kinds of dynamical theory which have been proposed to account for evolutionary dynamics, which I shall call "transformation theories" and "emergence theories", respectively.

1. Transformation Theories. Underlying all theories of this type is the view that every biological species can be represented as a point in an appropriate kind of space; this space can be coordinatized by a sufficiently comprehensive collection of morphological characteristics, each of which takes on a definite value for a particular species. Thus, such a theory proceeds directly from phenotypic characterizations. The transition from one species to another amounts then to a modification of the numerical values assumed by one or more of these morphological characteristics, causing the representative point to move in its space in such a way as to trace out what looks like a dynamical trajectory. The prototype of such a theory is of course that of D'ARCY THOMPSON (1); for a review of some other transformation theories see (3) loc. cit. These theories can be built at any biological level, and imply the familiar view that all organisms are in some sense "models" or analogs of one another.

Such theories are attractive because we can, according to Darwinian ideas on natural selection, view the trajectories as being minimizing curves for some kind of cost function, and we can obtain information about the selection processes involved (expressed in terms of the cost function) when the trajectories are known. But the main, and in fact fatal, difficulty in establishing such theories in dynamical and evolutionary terms is that the individual states ("species") which occur in such theories do not represent the states of a real system, and the trajectories in this state space are not parameterized by real time. Thus these transformation models differ from all other biodynamical models in a truly fundamental way, for in all other biodynamic theories the essential object of interest is a real system changing state in real time.

Another difficulty, of almost equal importance, is that it is assumed that all the morphological characteristics which co-ordinatize the space involved are known in advance, and that evolutionary developments are represented simply by a modification of the numerical values of these characteristics. This kind of view simply leaves <u>no room for the generation of novel functions</u> and structures which we recognize as the most fundamental and challenging aspect of evolution. That is, it cannot in principle account for evolutionary novelties; at best it can provide a descriptive and phenomenological treatment of evolutionary changes, which cannot be readily related to the dynamical processes occurring in real organisms and real populations.

2. Emergence Theories. The theories which fall under this heading find their prototype in the work of OPARIN (2) on the origin of life. The viewpoint taken here is essentially that of descriptive chemistry; we start with a population of simple elements of several different kinds, which can interact to generate more and more complex structures. These structures may have properties completely different in kind from those of the elements of which they are composed, but which can be understood in terms of the properties of the elements themselves and the rules governing their interaction.

In OPARIN's original theory, for example, we begin with a small number of different chemical constituents in a "soup"; these presumably interact with each other, in the presence of radiation, to form more and more complex organic compounds, until at some point the kind of organization which we recognize as life (e.g. a living cell) becomes inevitable. Such a theory has the virtue that the dynamical processes which occur in it are indeed changes of state of a single real system in real time, thus obviating the first basic difficulty noted in the transformation theories.

However, in formal terms, the "emergent" theories are no more emergent than the transformation theories are. For the <u>dynamical description</u> of the states involved always <u>remains at a low level</u> of system description; e.g. in den OPARIN theory it remains at the level of the values assigned to the concentrations of chemical species in the "soup". There is no point at which we can identify functional aggregates of greater complexity, expect by assigning a concentration number to them. That is, there is no way we can pass to descriptions of aggregates which function as "individuals" at a higher level (where by such an "individual" we would mean a pattern of temporal change in certain state variables which remain strongly correlated for long periods of time). The problem can be graphically stated as follows: how would one go about finding whether there was an amoeba in a pot filled with amoeba constituents, when the only information which can be obtained about the system is the change of concentration of the constituents in time? Obviously that information is not sufficient to find whether there is an amoeba in the pot or not; what is necessary is new information, pertaining to the properties of a functional individual at a higher level of system description. This is precisely what is missing from the "emergent" theories; they talk about emergence only at a single level of system activity and system description; they do not generate new state variables in an effective way.

Thus it is that both the transformation theories and the emergence theories fail to incorporate, let alone account for, the generation of evolutionary novelties. Now we may ask: is the situation really as dark as it seems? Is there in fact no way of understanding the generation of evolutionary novelties? The answer is that there is a way of understanding such processes, which was formerly much discussed, but which has resisted incorporation into dynamical models of any kind. This is the so-called <u>Principle of Function Change</u>, which asserts roughly that many different biological functions coexist at any time in any particular anatomical or physiological structure, and that evolutionary development has the effect of shiftingthe weight assigned to one function carried out by a structure at the expense of the others. Such a principle seemed to the early evolutionists the only way to account for structures like the eye, which is only of selective value when it can actually see; or the way in which an organ of equilibration like a swim-bladder in a fish can be transformed into an organ of respiration like a lung. Indeed, I might assert that no dynamical model of evolution can have any claim to validity unless it incorporates the idea of biological function, and something playing the role of a Principle of Function Change, in an essential way. And of course, neither the transformation theories nor the emergence theories do this; they are purely structural theories, and this is their main weakness. But it is a weakness common to almost all biodynamical models; the state variables in terms of which these systems are conventionally defined are almost entirely of a structural, and not of a functional, character. Thus I would regard this Principle as a kind of bellwether for the success of a dynamical theory of evolution.

A POSSIBLE SOLUTION

Now how can we go about actually constructing a good dynamical theory of evolutionary novelty? As PATTEE has so abundantly recognized, the kind of dynamical modelling necessary for a good theory of evolution (and of evolutionary novelties) must carry within itself the possibility of effectively generating, from the intrinsic dynamics of the system, meaningful new state variables appropriate for the description of new functions emerging as a result of the interaction between the evolving system and its environment. It must thus simultaneously be able to deal with state description at different organizational levels of the same system; i.e. it must be intrinsically hierarchical. It is this fact more than any other which has been a stumbling block to the construction of dynamical models of evolution, since most dynamical models in biology are trapped at a single organizational level (and hence certainly cannot account for the emergence of new properties at higher levels as a consequence of dynamical activities at the lower ones).

But there is perhaps a way of turning this defect into a virtue. For let us observe that it is actually meaningless to say that a system, by itself, is hierarchical. A system can look hierarchical to us if we interact with it in such a way that different state descriptions are appropriate to account for its behavior (5), or if it is placed in an environment with which it can interact at these several levels. Thus the hierarchical aspect is as much a consequence of the environment with which the system interacts as it is of the system itself; perhaps much more so, since as pointed out elsewhere (4), analogies between arbitrary systems are the rule rather than the exception. Thus the "origin of hierarchy", a problem to which PATTEE has devoted so much attention from the viewpoint of the origin of life, is essentially the problem of how to modify both the individual system, and the environment with which it interacts, so as to enhance the kind of hierarchical aspects which we subjectively see manifested in living systems.

Since, by our arguments on analogous systems (4), essentially all systems can be viewed as hierarchical, requiring only the proper environments to make their hierarchical aspects manifest, the problem of origin of hierarchy is thus reduced to successively modifying the system-environment interacting so as to increasingly amplify the hierarchical aspects of this interaction. This means in effect, successively changing the weights associated with the system-environment interactions so as to increase or magnify those which can appear to us as having a hierarchical character. This kind of shifting of weights in certain kinds of system-environment interactions should remind the reader of something we have already discussed; namely the Principle of Function Change, here manifesting itself at a lower level of organization than before. Its reappearance in this new context is in itself an interesting instance of the kind of system analogies involved in this discussion, and which permeate all of biology. The important novelty at this level, which distinguishes this kind of argument from the simple kind of parameter adjustment previously invoked to account for function change, is that, in effect, the parameter adjustment must take place in the system environment, or better, in the coupling between the system and its environment, rather than just in the system itself.

The next observation is that, in evolutionary dynamics, we always find the <u>emergence of new dynamical qualities at higher levels without any change in the dynamical qualities at the lower levels</u>. Let us discuss this peculiarity for a moment. Indeed, we find in all developmental and evolutionary processes, a curious population aspect. There is always an initial proliferative process, resulting in an increase in the dimensionality of the system at a particular dynamical level (i.e. the number of degrees of freedom of the system is increased) which adds no new <u>qualities</u>, but merely increases <u>quantities</u> or redundancy in a special way. For instance, in development we have initially a proliferation of cells, and in evolution initially a proliferation if individuals on which selection can act. The emergence of hierarchy in both cases arises from a reshuffling of this great quantity of new degrees of freedom so as to produce new correlations between them, ultimately manifesting themselves in the form of higher functional units, or new "individuals", with new functional properties. These "individuals", as noted above, retain the same lower-level qualities as before, (although, as PATTEE has emphasized, new correlations between these qualities may be imposed as a concequence of the higher-level organization).

This idea of <u>adding redundant degrees of freedom</u>, which can then vary independently to <u>produce essentially new qualities</u>, is an old one in evolution, but a new one to dynamical modelling. Tandem duplications of chromosomal segments or individual genes have long been regarded as one of the fundamental evolutionary raw materials, as have the appearance of organisms constructed of serially repeating body segments. What we wish to emphasize here is that this redundancy at one level of organization can manifest a hierarchical aspect as well; e.g. as when a population of gas molecules acts in concert to produce a qualitatively new effect (pressure), or when cooperative effects become manifest, as in phase transitions and crystallization.

There is a formal way of multiplying state variables, a mathematical equivalent of extending copies of the system in space, which may be of interest in this connection. As an example let us consider the TURING system (8), an individual cell of which is specified by a pair of dynamical equations

$$\frac{dx}{dt} = ax + by + S$$

$$\frac{dy}{dt} = cx + dy + S$$

in which x, y represent the concentrations of chemical species. N such cells, arranged in a ring, and interacting with neighboring cells by diffusion, satisfy the equations

$$\frac{dx_i}{dt} = ax_i + by_i + D_1 (x_{i-1} + x_{i+1} - 2x_i) + S$$

$$\frac{dy_i}{dt} = cx_i + dy_i + D_2 (y_{i-1} + y_{i+1} - 2y_i) + S$$

$$x_i, y_i \geq 0, \; x_i = x_{N+i}, \; i = 1, \ldots, N.$$

Each of these cells is specified by the same state variables as all the others, and in fact as the whole system; no new qualities have been introduced, but the quantities have been modified in such a way that the dimensionality of the system has been raised. In effect the system has been extended in space, with the index of summation i representing a (discrete) variable of extension. Indeed, if we add up the equations, we get back to the original single TURING cell, when we put

$$\sum_{i=1}^{N} x_i = x$$

$$\sum_{i=1}^{N} y_i = y$$

If we looked only at the qualities x, y for the entire ring, it would look to us as if we had a single system, lumped in space, moving to an asymptotically stable equilibrium; but when we segment or partition the system so as to redundantly multiply the number of degrees of freedom without introducing any new qualities, we can generate new and very complicated spatial relationships between these qualities, even of an essentially hierarchical kind, within that same dynamical framework.

Indeed, quite generally, given even the single dynamical equation

$$\frac{dx}{dt} = f(x)$$

if we introduce new observables x_i, $i = 1, \ldots, N$ such that

$$x(t) = x_1(t) + x_2(t) + \ldots + x_N(t)$$

then <u>any</u> dynamical system of the form

$$\frac{dx_i}{dt} = \varphi_i(x_i) + g_i(x_1, \ldots, x_N) \quad i = 1, \ldots, N$$

where

$$\sum_{i=1}^{N} g_i = 0, \quad \sum_{i=1}^{N} \varphi_i(x_i) = f(x)$$

will have the above characteristics. As noted above, this is the mathematical equivalent for introducing a spatial redundancy by serially repeating copies of the system in space, and allowing for interaction between the redundant systems consistent with the overall dynamics of the system. If we now allow the system to interact with another system, which sees the x_i instead of seeing x, we will have made much progress.

A process like this can indeed "find the amoeba in the pot" if it is developed appropriately.

It is hoped that the foregoing considerations may have shed some light on the underlying reasons for the difficulties arising from emergent processes in dynamical models of evolutionary processes, and perhaps thereby have contributed something useful towards the resolution of these difficulties.

References

1. D'ARCY Thompson: On Growth & Form, Cambridge 1924.
2. OPARIN, A.I.: The Origin of Life, MacMillan 1938.
3. ROSEN, R.: Optimality Principles in Biology, Butterworth (1967).
4. ROSEN, R.: On Analogous Systems, Bull. Math. Biophy., $\underline{30}$, 481 (1968).
5. ROSEN, R.: Hierarchical Organization in Biological Systems. In: Hierarchical Structures, Whyte, Wilson & Wilson (eds.) Elsevier, p.179 (1969).
6. ROSEN, R.: Dynamical Systems in Biology, Wiley, 1970, in press.
7. STAHL, W.F.: Physiological Similitarity and Modelling, 1967.
8. TURING, A.M.: The Chemical Basis of Morphogenesis, Phil. Trans. Roy. Soc., $\underline{37}$, 1953.

DISCUSSION

BREMERMANN:
ROSEN says: .... the individual states ("species") which occur in such theories do not represent the states of a real system, and the trajectories in this state space are not parametrized by real time". – What is meant by "real systems"? Are hypothetical thermodynamical or gravitational systems (as appear in physics text books) more "real"? And how can one say it is not parameterized by real time? Isn't that up to the modeller? OPARIN's theory of the origin of life is mentioned. Currently Manfred EIGEN is publishing a modern version of it that seems highly relevant to the issues of this paper. – It seems to me that the amoeba very carely appears in the "pot of constituents". It is the product of a very long process of chemical and primitive evolution, perhaps as long as that which led from amoeba to man.

ROSEN:
When I stated that in transformation theories like that of THOMPSON the individual states were not the states of "real" systems, changing in real time, I meant something like the following. Suppose we consider the evolutionary transformation from Eohippus to Equus. This transformation can be idealized as a deformation, á la THOMPSON, which can in turn be represented as a trajectory in a suitable state space. Certain points on this trajectory can be identified with known fossil or modern forms, such as "Eohippus", "Mesohippus", etc.; the other points with conjectured intermediate forms. But the point labelled, e.g., "Eohippus" does not refer to any actual individual organism; it rather refers to some average or idealized type drawn by composition from all specimens to which the name "Eohippus" is given. The type of "Eohippus" remained constant (or approximately so) for an indefinite (and certainly rather long) time before a typical modification of form became apparent. Thus there is no question of individual organisms, during their lifetimes, tracing out the transformation of form, but rather a succession of abstract "types" which are ordered into "earlier" and "later" according to an arbitrary temporal scale (and hence not "real time"). In development models, on the other hand, we do deal with a single system, changing state in real time. This is the distinction I intended, and I think it is an important one. Incidentally, an interesting variant of the transformation theories was suggested by a number of biologists about thirty years ago (cf. the discussion in Chapter 6 of my book, "Optimality Principles in Biology", Butterworth , 1967), where they argued that the entities being transformed were not isolated "adult types", as in THOMPSON's theory, but rather whole developmental histories. This led to NEEDHAM's (and others') idea that all organisms were spatio-temporal models of each other, an idea which independently reappeared in another form in RASHEVSKY's relational biology. – I didn't mean to raise the question of whether a real

amoeba would (or would not) appear in a pot of amoeba constituents. I was rather dealing with the question of whether we could tell whether or not an actual amoeba was present in such a pot, given that all we can monitor or detect were such constituents. The problem I raised is not thus an evolutionary problem, but rather a problem connected with the <u>observation</u> or detection of organized individuals within the context of a circumscribed set of lower-level observation techniques. And as I tried to indicate in the body of my paper, the capability of detecting such individuals is a crucial aspect of certain kinds of evolutionary formalisms.

PATTEE:
ROSEN's idea of adding degrees of freedom by considering collections of similar units whose dynamics are initially redundant, but which generate new correlations and hence new quantities may be an important part of the origin of hierarchical controls. However, I am still puzzled by the origin of the actual constraints which establish and execute the correlations or hierarchical rules. I want to emphasize the necessity for real machinery to perform functions. In living systems these machines begin with single molecules – the enzymes and nucleic acids – which normally function only in highly coordinated groups involving several levels of hierarchical control.
All concepts of dynamical systems, including the concept of physical laws, involve two distinct types of construct – a set of rules and a set of quantities obeying these rules. The "rules" we elevate to the level of physical laws differ from all other rules, such as the rules of a game or an automaton, by being incorporeal and inexorable. By incorporeal I mean that physical laws need no observable device to assure their execution. All other rules require a real, physical device or mechanism to exist in any objective sense. By inexorable I mean that physical laws are strictly deterministic in the sense that it is inconceivable that they be disobeyed or in error. If it should appear that a law does not fit our observations we say that either our observations are in error or that we have not stated the dynamics of the law correctly. More directly we could say that we do not think of the world as being in error – only our measurements and our descriptions of the world can be in error. – When ROSEN speaks of one system "seeing the x_i instead of the x" this obscures the fact that "seeing" is not a physical law but must result from a special physical constraint which classifies interactions into what is seen and what is not. All processes of classification, which may be disguised under many names – decomposition, pattern recognition, feature extraction, weighting, selecting, measuring, etc. – require a real physical embodiment to perform what we can too easily describe as an abstract process. – The difficulty, as I have emphasized in my chapter in this volume, is that the very concept of a constraint in physics requires an alternative description to the dynamical description of the underlying units. Thus an objective hierarchical control must effectively generate a description of the motion it constrains. The genotype of the organism is a description of the organism, but any description must have a set of rules for writing and reading. The genetic code and translation mechanisms supply these rules in the cell, however, they too are constraints and must themselves be described by the gene. This is simply a molecular form of the chicken–egg paradox. My point is that all hierarchical control origins suffer from this type of paradox. Hierarchical rules are created only by constraints, and constraints are only alternative descriptions of their underlying dynamics. But we have no meaning for description expect as generated by a set of constraints forming some kind of language structure. In some sense, therefore, objective hierarchical constraints must contain their own description. – Of course we have many examples of subjective hierarchies created by our own choice of description. For example, we usually choose to describe a gas by its pressure rather than the position and momenta of its molecules. But notice that there is no effect of our concept of pressure on the actual motions of the molecules. That is, the pressure does not act as a constraint. On the other hand, if there should exist a real device or mechanism which constrains the motion of molecules by measuring the pressure (a regulator valve), or which constrains the pressure of the gas by measuring the motion of the molecules (a MAXWELL demon), then indeed we have an objective hierarchical control system. But clearly this type of constraint does not arise simply from the integration of dynamical equations. – The origin of life problem is made even more obscure, I believe, because this type of non-integrable constraint that characterizes living systems is often embodied in sets of single molecules. Thus we might expect the underlying description to be quantum dynami-

cal. But quantum dynamics differs from classical dynamics because the quantities which obey the dynamics are not directly observable. It is necessary to use classical constraints to make quantum mechanical measurements. This poses a serious conceptual problem for which there is still no lucid explanation even in the context of ordinary physics. Yet I do not see how we can evade the problem at the origin of life and still claim to understand the physical basis of life.

ROSEN:
As PATTEE indicates, there are enormous gaps which obstruct all but the crudest kinds of qualitative suggestions for how hierarchically organized control systems can be generated. I have offered several such suggestions in my paper, and elsewhere, as have many other people; but there is as yet nothing like a theory to bridge these gaps. Perhaps reflections of the following kind may be of some value in this connection. - PATTEE states that physical laws are _incorporeal_. This assertion is certainly correct with respect to our own observations of physical systems. But I would like to suggest that interacting physical systems are themselves continually observing each other, and indeed, that their interaction can be looked at as the simultaneous and continuous measurement of (observables of) each system by all of the others. In dealing with such general interactions, physical theory is surprisingly of little help, being restricted to special systems (conservative ones) of limited biological interest, and beyond this to the limiting case of infinitely weak interactions (perturbation theory). In general, physics gives us no insight into how to write down equations of motion of a composite system, even supposing we know the equations of motion of the component subsystems. This being so, the behavior of aggregates of initially identical systems, which can interact (i.e., measure) with each other through many observables is, in the present state of knowledge, quite unpredictable, and can be indefinitely rich. - Thus I would say that "seeing" actually involves physical law in an essential way, and beyond this, that all physical laws are assertions about how systems "see" or "measure" each other; i.e., how the values of observables defined on the states of one system are reflected in values of observables of other systems with which the first interacts. - Another basic difficulty involves the fact that we have one language for describing states and quite a different one (and a far less well-developed one) for describing mappings or processes. If you simply ask how one "observes" a mapping process, or how one attaches "observables" to such a process, you will immediately see the dichotomy between the description of system states and system dynamics. Since "measurements" or "observations" are processes, this dichotomy becomes particularly acute when referred to precisely the things we need to know in order to study system interactions. Essentially, the gap between state description and process descriptions is another form of PATTEE's paradox, and it seems that until this gap is bridged, no quantitative progress in these areas is possible at all.

FONG:
Perhaps the difficulty of the transformation theory may be overcome by incorporating in the coordinate space all morphological characteristics of all living species. The coordinates of a point representing a given species will thus be mostly zero (for characteristics not belonging to the given species) except a few (for characteristics of the given species). The emergence of a new characteristic may thus be represented by a rotation of the state vector in the multi-dimensional space in which some coordinate value changes from zero to non-zero. Nevertheless there are evolutional changes, such as the evolving of the "fins" of marine mammals from the legs, that cannot be represented this way. For such cases, the state of a "limb" may be represented by a superposition of two states, the leg and the fin, in much the same way as quantum mechanical states are superimposed. The discontinuous evolution of a leg to a fin is thus represented by a continuous change of the superposition coefficients from $(1,0)$ to $(0,1)$. The mathematics to handle this kind of situation has been well developed in quantum mechanics. LANDÉ has made the point that the superposition of quantum states is the way by which physical discontinuity may be handled mathematically in a continuous way. The idea may be applied to other fields. The concept of superposition is not necessarily abstract and mathematical. Considering the fact that man's limbs are also used for swimming and seal's fins are also used for walking it is quite obvious that the state of the limb may be represented by a superposition of two states.

ROSEN:
The idea of superposition certainly has attractive features for the modelling of function change. The basic question remains, however; what are the entities which are to be superposed? Perhaps these entities should be whole physiological or developmental models; as I remarked in my paper, we typically abstract out the multiplicity of function in the process of constructing the model, thereby forcing the model to be in a "pure state" with respect to particular biological functions.

GÜNTHER:
In connection with the Principle of Functional Change, I would like to draw the attention to the experiments on cold receptors of the cat tongue put foward by HENSEL & ZOTTERMANN (Acta physiol. scand. 23: 291-319, 1951). Although the integrated frequency recorded from a single fiber showed no correlation with temperature; only a limited correlation was found on the 4-5 fiber preparation, restricted to the range of 40° C to 25° C. Nevertheless, a linear correlation with temperature was observed when action potentials from 10 or 20 fibers were recorded simultaneously. This so-called "Thermometer-Function" seems to be the result of the redundancy of cold-receptor activities, transmitted through several afferent pathways at the same time.

GABEL:
ROSEN's comments of "transformation" theories and "emergence" theories for evolution and biogenesis are very apropos as a criticism of the current consensus of opinion subscribed to by many contemporary cosmogonists. These theories are indeed purely structural theories and ROSEN's assertion that "no dynamical model of evolution can have any claim to validity unless it incorporates the idea of biological function and something playing the role of a "Principle of Function Change" is worth repeating for emphasis. A theory for biogenesis which accounts for the interrelationship of function and structure was proposed by me in 1965. Although this theory is consistent with geophysical reality, it, unfortunately, has not been widely acclaimed. My paper of this symposium is a generalized expansion of this theory which is based on the excitability property of matter. In its conception this theory would have been classified as geophysical emergence. In its present form, I am not sure how ROSEN would classify it. Most aspects of it can be treated from either the transformation or emergence point of view. - The maintenance of a small population of novel functions and structures is, in itself, of potential survival benefit to life on a species level and also on the scale of global optimization. A particle matrix endued with sufficiently structuralized morphological or metabolic characteristics will have duration. The survival benefit of the incorporation of novelties into social structures as well as inanimate objects has been pointed out by W.V. SMITH (1970). The proliferation of biological populations which always precedes developmental and evolutionary processes is analogous to the fragmentation and restructuralization which occurs during chemical evolution.

COULTER:
ROSEN's critical analysis of the inability of transformation and emergence theories to account for the emergence of new dynamical qualities in evolution is cogent, and his proposal for constructing a dynamical theory of evolutionary novelty is intriguing. It occurs to me that a mathematical theory of synergic systems might be applicable here. This implies the existence of two mechanisms (or system properties): (1) a decision-making component with the capacity to select or filter some of the interactions (those which are synergic), and (2) a mechanism for assigning worth-value to some set of properties of the system as a whole (i.e., the organism-environment considered as a system). Such a "global utility function" could measure, for example, the "fitness of the environment" of which HENDERSON wrote, as well as internal interactions. None of this, of course, implies any Lamarckian adaptations or vitalistic entelechies, since synergic systems are simulable on a computer.

REANNEY:
Another way of approaching the problem of the origin of metabolic novelties comes to mind. BRILLOUIN (Science & Information Theory, Acad. Pr. 1956) has suggested an approach based on

information theory. One can use the data of molecular genetics. The number of bases in the DNA of Escherichia coli is 10^7. By interpolation into the information equation

$$I = 1.38 \times 10^{-16} \ln P_o \quad \text{where } P_o \text{ here} = 4^{10^7}$$

we get a figure of 2×10 ergs/'C representing the structural negentropy of the organism (see LWOFF "Biological Order" MIT Press 1962). As the <u>quantity</u> of DNA increases during evolution so must the numerical value of the structural negentropy to the organism. But here we run up against the difficulty that negentropy, thus defined, does not take into account the evolutionary past, which fits one particular base sequence optimally to one environment (this is a loose statement since degeneracy of the code and the occurrence of silent mutations allow a variety of different base sequences to generate the same phenotype). Any sequence of 10^7 bases has the same value for structural negentropy but biologically only a limited number allow the system to survive in the codon. I think one must do what ROSEN suggests here and study "the coupling between the system and its environment rather than just the system itself". The problem is that while one can perhaps define the "quantity" of information mathematically, the "quality" of information cannot be specified.

LOCKER:
Your distinction between transformation theory and emergence theory is fundamental and absolutely well justified; also your showing forth the deficiencies inherent in both theories rendering them unable to really explain the origin of evolutionary novelties or even of life itself. Because it is not easy to define reality I would not feel necessary to grasp the essence of a species as a non-real system. Although a species is indeed more abstract than an organism, both, organism and species, share reality in that they exhibit existence. The other difficulty mentioned by you, namely that in the transformation theory in its present garb the morphological characteristics should be known in advance, touches upon a point which also THOM has referred to by drawing attention to the logical difficulties which arise wherever one is tempted to extrapolate from the individual organismic development to the evolution of the biosphere. The Principle of Functional Change you are quoting is a fairly good starting point for avoiding these difficulties in that it delivers a conceptual framework wherein the different functions potentially coexist since only by this very fact is implied the possibility for a true generation of novelties. This, by all means, must occur through a transformation of something potentially pre-existing into an actually existing one. In cogitating about the mode by which you (as well as PATTEE) were able to make the addition of redundant quantities produce new qualities, it appears that a fruitful application of an idea has been made, domiciled in philosophy and going back to HEGEL, namely the revulsion of the quantitative into the qualitative. It deserves merit that you have found a way to mathematically formalize such a revulsion by exemplifying with the TURING system the generation of new state variables. Probably, that on the basis of any occurrence of novelty as being tantamount to entry into existence (i.e. becoming perceptible) of something pre-existing (i.e. mentally conceivable), - intimated also by the recent advances of topology with reference to structural stability, and in connexion with a mathematically well founded description of the quantitative to qualitative transition, - a true theory of the origin of life could spring out. I am also attired by your new-fashioned view of hierarchies as resulting from the interaction between system and environment, which idea introduces an objectivization of the constructive element for generation of hierarchies being connected with the subject-like nature of the system.

COHEN: see page 35

ROSEN:
COHEN makes an interesting point in connection with his statement that "evolutionary flexibility is itself the result of selections which have operated in the history of the species". He con-

cludes, if I understand him correctly, that organisms which have evolved in continuously fluctuating environments (such as many bacteria) would exhibit physiologies consisting of rather autonomous, loosely coupled subsystems; this would serve both as a buffer against environmental fluctuations and as a source of genetic variability. On the other hand, organisms developing in relatively constant environments would tend to have a much more highly integrated physiology, without many obviously semi-autonomous subsystems. This might help account for why molecular biology, which proceeds by means of decomposition into subsystems of a special type, has had so much success with bacterial systems, and conspicuously less success with other kinds of systems.

Circular Nucleic Acids in Evolution

D. C. Reanney

Abstract +

Some of the selective advantages conferred upon DNA by cyclication are enumerated. These may explain the occurrence of circular DNA in bacteria capable of sexual transfer, in episomes capable of stable integration into the host genome, in oncogenic viruses and in small DNA phages. These advantages, together with other considerations, suggest that cyclic polynucleotides played a critical role in early evolution.

Reference

REANNEY, D.C., RALPH, R.K.: J. Theor. Biol. 15, 41 (1967); 21, 217 (1968).

DISCUSSION

GABEL:
HALMANN has given experimental evidence which questions the significance of condensing agents such as cyanimides on the primitive earth. The experiments quoted by REANNEY & RALPH (1968) are operative only under acidic conditions. There is reason to believe that the primordial seas and atmosphere were neither acidic nor strongly alkaline. REANNEY's argument on this point would be considerably strengthened if he substituted a more likely primordial condensing agent which also happens to be itself a biopolymer, namely, inorganic polyphosphate. Thermally produced polyphosphate mixtures have been shown to contain a broad range of cyclic rings. Cyclic polyphosphate (metaphosphates) could perhaps have served as the template and condensing agent of cyclic polynucleotides. Inorganic polyphosphates have been reported to be intimately associated at the structural level with polynucleotides. Metabolic evidence from contemporary microorganisms also suggests that inorganic polyphosphate may have been the evolutionary precursor of nucleoside phosphates.

REANNY:
GABEL makes a good and valid point in drawing attention to the possible role of inorganic polyphosphate as a primitive condensing agent. Random condensation of polynucleotides usually produces a small fraction of closed circular molecules so that cyclic nucleic acids would have been generated by most of the currently postulated agents: but cyclisation may have been aided by the presence of cyclic structures in polyphosphate, as GABEL suggests.

+ Full paper will presumably appear in BG & Th.

Physiological Time and Its Evolution
B. Günther

Abstract
The phyletic increase in body size of invertebrates and vertebrates is correlated with a parallel reduction of the metabolic rate per unit mass and the frequencies of several endogenous oscillatory phenomena. - Chronobiological processes are discussed in accordance with the theories of mechanical and biological similarities. The concept of "operational time" leads to quantitative predictions of the allometric exponents of numerous time functions, which are in agreement with the experimental findings. - Synchronization mechanisms probably of neuro-humoral nature are postulated in order to correlate different functions at the cellular, organic and organismic level.

In accordance with COPE's law (1885) there is a persistent and widespread tendency for body size in animals-invertebrates and vertebrates - to increase during their phylogeny (16). This evolutionary trend of body weight (W) increments, must have been associated with adaptive changes of the metabolic rate (M), expressed for instance as oxygen uptake per organism per hour.

The relationship between body size and metabolic rate will follow one of these three possibilities:

1. The total oxygen uptake increases linearly with body size or with the total mass of cells pertaining to an organism; this relationship could be defined as $M \propto W^{1.0}$;

2. The metabolic rate is proportional to the body surface or to any other significant exchange surface, as for instance, the absorption area of the intestine, the gas transfer surface at the lung alveoli, or the heat dissipation area at the body surface; in these particular cases the power law reads as follows: $M \propto W^{0.66}$.

3. The correlation between respiratory rate or heat production and body weight assumes an intermediate value ($W^{0.66} < W^b < W^{1.0}$).

It is worth mentioning that VON BERTALANFFY (2) established a close correlation between these three alternatives (metabolism vs. body weight) and the corresponding growth types.

In any case, the correlation between body mass and oxygen uptake could be extrapolated eventually to the chronological events within the organisms. As it is well known, the phylogenetic and ontogenetic increase in body size is generally associated with a progressive reduction of the metabolic rate per unit mass - the so called "Stoffwechselreduktion" of LEHMANN - which influences the turnover rate at the cellular level and consequently could modify the biological timing processes.

Furthermore, we accept that for the evolution of life on earth every biological system must have followed the laws of physics.

Based on these assumptions, it seems licit to postulate that certain mechanical devices were eventually employed to measure "time", and consequently "mechanical" similarity criteria should be applied when comparing chronological events in small and large animals.

For this specific purpose, one organism can be defined as the "prototype" (subindex 1) and any other as the "model" (subindex 2), where the mass ratio between both leads to $M_1 / M_2 = \mu$, the corresponding lenght ratio to $L_1 / L_2 = \lambda$ and the time ratio to $T_1 / T_2 = \tau$.

I. PHYSICAL SIMILARITIES

The "mechanical" similarity principle of NEWTON implies the acceptance of the following two postulates:

First postulate: Since prototype and model should be made of the same materials, it is necessary to admit that the density (ρ) of both structures is constant (ρ) = (ML^{-3}), and consequently $\mu \cdot \lambda^{-3} = 1.0$, so that $\mu = \lambda^3$ or $\lambda = \mu^{1/3}$.

Second postulate: We must also assume that the acceleration of gravity (g) is constant (g) = (LT^{-2}), which means that $\lambda \cdot \tau^{-2} = 1.0$ and therefore $\lambda = \tau^2$ or $\tau = \lambda^{1/2}$.

Any biological function (Q), defined dimensionally in accordance with the MLT - system, can be written as follows:
a) for the "prototype"

$$Q_1 = M_1^\alpha \cdot L_1^\beta \cdot T_1^\gamma$$

b) for the "model"

$$Q_2 = M_2^\alpha \cdot L_2^\beta \cdot T_2^\gamma$$

By means of NEWTON's reduction coefficient ($\chi = Q_1/Q_2$), it is possible to establish a formal relationship between the functions of the prototype (Q_1) with those of the model (Q_2), simply by dividing the above mentioned two equations:

(1) $$\chi = \mu^\alpha \cdot \lambda^\beta \cdot \tau^\gamma$$

When the equivalent values, deduced from the two postulates ($\mu = \lambda^3$ and $\tau = \lambda^{1/2}$), are introduced into Eq.1, we obtain the general equation for "mechanical" similarities:

(2) $$\chi = \lambda^{3\alpha + \beta + 0.5\gamma}$$

As an example, let us calculate the "reduced exponent" of the length ratio (λ) for a periodic phenomenon, for instance, the duration of one oscillation of a simple pendulum, whose length is L. The physical dimension of a "period" is ($M^0L^0T^1$) and therefore the numerical values of the exponents are for $\alpha = 0$, for $\beta = 0$ and for $\gamma = 1$. From Eq.2 we obtain a reduction coefficient $\chi = \lambda^{0.5}$, which is equivalent to the classical equation for the duration of one period of a pendulum:

$$T = 2\pi \sqrt{\frac{L}{g}}$$

where:
L = length of the pendulum, and
g = acceleration of gravity.

In biology, particularly when we are dealing with interspecies comparisons, the length ratio (λ) can be replaced conveniently by the mass ratio (μ) or even by the weight ratio ($\omega = W_1/W_2$). Thus, Eq. 2 may be written as follows:

$$\text{(3)} \qquad \chi = \omega^{\alpha + \frac{\beta}{3} + \frac{\gamma}{6}}$$

because $\lambda = \omega^{1/3}$

The power function (Eq. 3) for a periodic phenomenon whose dimensions is (T), yields a reduction coefficient $\chi = \omega^{0.16}$ and for any frequency (T^{-1}) we have therefore $\chi = \omega^{-0.16}$.

Now, it seems necessary to establish the correlation between these theoretically deduced power functions and the empirical "allometric" equations, which after HUXLEY (12) may be written as:

$$\text{(4)} \qquad y = a \cdot W^b$$

and taking logarithms of both sides of Eq. (4) we obtain:

$$\log y = \log a + b \log W$$

where:
y = any biological function defined in accordance with the MLT-system;
a = empirical parameter which is equivalent to y when W = 1.0;
b = reduced exponent; and
W = body weight expressed in g or Kg.

From the experimental data concerning heart rate and respiratory frequency in mammals of different body size (from the 2.5 g shrew up to the 4 tons elephant) it follows that the reduced exponents (b) for these frequencies are -0.27 and -0.28 -respectively. Since both exponents were obtained after the statistical treatment of numerous experimental findings, we are sure that these exponents differ significantly from the theoretically predicted values (b = -0.16). In consequence, we must conclude that the "mechanical" similarity criterion is unable to predict correctly the "reduced" exponent (b) for biological time functions when they are expressed as allometric equations.

II. BIOLOGICAL SIMILARITIES

In view of the discrepancy between the mechanical similarity principle and the experimental findings, LAMBERT & TEISSIER (14) proposed in 1927 a new theory of biological similarities. These authors retained the first postulate of the mechanical similarity with the organism density (ρ) as a constant, but replaced the second postulate by an a priori assumption in the sense that $\tau = \lambda$. The introduction of both these equivalences into Eq. 1 yields the general equation for "biological" similarities:

$$\text{(5)} \qquad \chi = \lambda^{3\alpha + \beta + \gamma}$$

which also can be expressed as a function of the body weight ratio (ω):

(6) $$\chi = \omega^{\alpha + \frac{\beta}{3} + \frac{\gamma}{3}}$$

The reduced exponent (Eq. 6) for a period (T) assumes now a value of 0.33, whereas for any frequency (T^{-1}) we obtain $b = -0.33$.

In consequence, the new theory represents a real improvement, since the difference between the theory ($b = 0.33$) and the experiment ($b = 0.27$) is significantly smaller, when compared with the classical rules valid for mechanical similarities.

Let us calculate now the theoretical value of the exponent (b) for another function of time, which has been studied quantitatively for many years, i.e., the metabolic rate (H) of homeothermic animals. The physical dimension of metabolism (H) is equivalent to a "power" function (H) = (ML^2T^{-3}). The reduced exponent for metabolic functions is $b = 1.16$ in accordance with Eq. 3 (mechanical similarity), whereas the value is $b = 0.66$ after Eq. 6 (biological similarity); the commonly accepted empirical value is $b = 0.734$ for the basal metabolic rate of homeothermic animals.

III. THE "OPERATIONAL TIME" IN BIOLOGY

In order to minimize the differences between the theoretical predictions and the experimental findings, the concept of "operational time" was introduced (10), in the sense that a distinction should be made between the "physical" dimension of time (T) in agreement with the classical MLT-system, and the "operational" criterion for time which depends exclusively from the circumstance whether or not a "clock" is employed during the experimental measurements.

For the above mentioned purpose it was necessary to calculate first the exact difference between the theoretical and the empirical values for a biological "period" (0.333 - 0.27 = 0.063), and secondly, the difference for the "metabolic rate" (0.734 - 0.666 = 0.068), yielding a mean value of 0.065, when both cases are considered together. The next step was to separate the original formulation of LAMBERT & TEISSIER (14) from this new "correction" factor:

(7) $$\chi = \omega^{\alpha + \frac{\beta}{3} + \frac{\gamma}{3}} \cdot \omega^{-0.065 \Gamma}$$

where Γ is the operational time exponent, which may attain the following values:
$\Gamma = 1.0$ for any "period", measured experimentally by means of a clock.
$\Gamma = -1$ when we are dealing with a "frequency".
$\Gamma = 0$ when no clock is used during the experiment.

Thus, if we apply Eq. 7 in order to calculate the reduced exponent (b) of a "periodic" phenomenon in biology, we obtain:

$$T = \omega^{0 + 0 + 0.33} \cdot \omega^{-0.065(1)} = \omega^{0.27}$$

Conversely, for any "frequency" the predicted value is $b = -0.27$.

It is interesting to note that Eq. 7 has been applied to calculate the reduced exponent (b) of numerous other biological functions, with the same satisfactory result (10).

IV. THE DIFFERENCES BETWEEN "PHYSICAL" AND "PHYSIOLOGICAL" TIME

In accordance with classical physics "time" is a continuous forward flowing variable (EDDINGTON's time-arrow) with an infinitely small present (dt) which separates the past from the future. Another characteristic is that the direction of time is irrelevant, since physical equations are valid irrespective of their sign (dt or -dt).

The above mentioned physical criteria regarding time are ordinarily extrapolated to the biological realm. Nevertheless, it seems reasonable that any rhythmic phenomenon, which can be observed in organisms, eventually could be used as an index for a "biological time" scale. In previous papers (6, 7, 8, 9) a quantitative analysis of many periodic processes was made and discussed in the light of the theory of biological similarities.

The impossibility to separate the biological from the physical oscillatory processes should be emphasized, since the "mechanical" similarity is only a limiting condition for biological similarities. WEVER (18) has shown recently that a wide range of oscillatory phenomena can be described by the same differential equation, where one extreme corresponds to the physical "pendulum" and the other to the so called "relaxation oscillation". The type of oscillation only depends on the energy exchange between the oscillatory system and the environment; in one case we are dealing with a typical "mechanical" phenomenon (pendulum) and in the other (relaxation oscillation) with a characteristic "biological" periodicity.

However, the problem of biological chronometry is related to endogenous periodicities at different levels of organization, i.e., the cellular, organic and organismic levels. In addition, it seems necessary to postulate the existence of some kind of "synchronization mechanism" among the different internal clocks, which eventually could be mediated by neuroendocrine transmitters.

For instance, we can consider cardiac activity as a biological clock, whose regular succession of events -excitation, transmission and contraction- is a kind of timing device. Each cardiac cycle represents a discrete time unit, which may have a different duration - when measured by means of a physical clock - but during each period we observe always the same sequence of events, as for instance: several heart sounds, characteristic waves of the electrocardiogram, some singular points of the pressure - volume diagram, and many other oscillatory phenomena of physiological or biochemical nature. When we are dealing with the same model of animal, mammals for instance, the chronological evolution of each cardiac cycle is almost identical in successive cardiac periods and is independent of body size. Only the "duration" of each cardiac cycle is a function of body mass (Fig. 1), as has been postulated by the theory of biological similarities and confirmed by the statistical analysis of heart rate versus body weight data. In both cases the reduced exponents of the allometric equations are the same (b = 0.27).

As mentioned above, synchronization processes should be postulated among the different levels of organization, i.e., the pacemaker activity at the cellular level (sino-auricular node of the mammalian heart), the mechanical activity of the muscular pump (pressure-volume phenomena), and the energy and oxygen needs of all cells, which are supplied through the systemic circulation. As it is well known, in mammals the metabolic rate per unit body mass ($W^{0.734}/W^{1.00} = W^{-0.27}$) can be defined by means of the same power function as the heart rate or the respiratory rate vs. body weight ($W^{-0.27}$), relationships which are illustrated in Fig. 2.

The postulated functional correlation among the cellular, the organic and the organismic levels is probably not restricted to the relationship between heart's activity and metabolic rate, but must also be valid for the control of the respiratory rate, the gut beat duration, the sexual cycle and the life span (10), as well as for many other periodic phenomena within individual cells (4), as for instance the concentration of certain metabolites and cellular constituents (5).

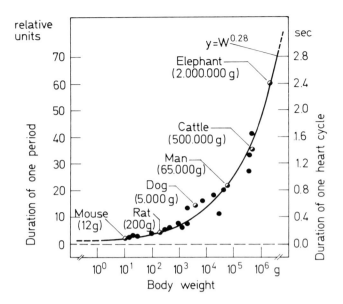

Fig. 1. Durating of one cardiac cycle in mammals of different size (8). Left ordinate: Time in relative units. Right ordinate: Time in seconds. Abscissa: Body weight in gram (logarithmic scale)

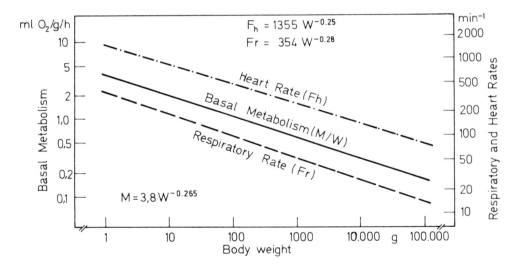

Fig. 2. Relationship between metabolic rate per unit mass (15), respiratory rate (1), cardiac frequency (17), and body weight (g) of mammals; in double logarithmic scales

Due to these hypothetical synchronization mechanisms, we may assume that any change of the metabolic rate should influence the heart rate, as for instance during excercise, hyperthyroidism, hyperthermia or fever, and finally due to the action of certain drugs (11).

Furthermore, we may extrapolate these general conclusions to the psychological level, due to the fact that the "time experience" is of finite and discontinuous nature, as has been found by EFRON (4) and KEIDEL et al. (13), among others.

Thus, at all levels of biological organization we can observe that the ontogenetic and phylogenetic increase in size (COPE's law) is directly associated with an increase of the duration of all periodic phenomena, a general trend which can be formulated by means of the power law for biological time: $T_B = aW^{0.27}$.

In conclusion, the evolutionary increase in body mass (W) is correlated with a reduction of the metabolic rate (M) per unit mass ($M/W = W^{-0.27}$) and a parallel frequency decrease of all periodic phenomena ($F_B = W^{-0.27}$).

ACKNOWLEDGEMENTS

I wish to express my appreciation to Dr. Federico LEIGHTON for reading the manuscript and for his valuable suggestions.

References

1. ADOLPH, E.F.: Science, 109, 579 (1949).
2. BERTALANFFY VON, L.: Amer. Natural., 85, 111 (1951).
3. BETZ, A.: In: Quantitative Biology of Metabolism (A. Locker, Ed.) p. 205, Berlin: Springer-Verlag, 1968.
4. EFRON, R.: In: Interdisciplinary Perspectives of Time (E.M. Weyer, Ed.), Ann. N.Y. Acad. Sci., 138/2, 713 (1967).
5. GOODWIN, B.C.: In: Interdisciplinary Perspectives of Time (E.M. Weyer, Ed.), Ann. New York Acad. Sci., 138/2, 748 (1967).
6. GÜNTHER, B. and GUERRA, E.: Acta Physiol. Latinoamer., 5, 169 (1955).
7. GÜNTHER, B. and LEÓN DE LA BARRA, B.: Bull. Math. Biophys., 28, 91 (1966).
8. GÜNTHER, B. and LEÓN DE LA BARRA, B.: Acta Physiol. Latinoamer., 16, 221 (1966).
9. GÜNTHER, B. and LEÓN DE LA BARRA, B.: J. Theoret. Biol., 13, 48 (1966).
10. GÜNTHER, B. and MARTINOYA, C.: J. Theoret. Biol., 20, 107 (1968).
11. HOAGLAND, H.: In: The Voices of Time (J.T. Fraser, Ed.), p.312, New York: George Braziller, 1966.
12. HUXLEY, J.S.: The Problem of Relative Growth, London: Methuen, 1932.
13. KEIDEL, W.D., BREUING, G. and WIEGAND, P.: Ann. Acad. Sci. Fennicae, Helsinki, Ser. A., V. Med., 143, 1 (1969).
14. LAMBERT, R. and TEISSIER, G.: Ann. Physiol. Physiocochim. Biol., 3, 212 (1927).
15. MACMILLEN, R.E. and NELSON, J.E., Amer. J. Physiol., 217, 1246 (1969).
16. NEWELL, N.D.: Evolution, 3, 103 (1949).
17. STAHL, W.R.: J. Appl. Physiol., 22, 453 (1967).
18. WEVER, R.: In: Circadian Clocks (J. Aschoff, Ed.), Amsterdam: North Holland Publ. Co., 1965.

DISCUSSION

FONG:

The theory of mechanical similarity can be eliminated on logical ground. The assumption (2) that the constancy of gravitational acceleration g leads to $\tau = \lambda^{1/2}$ is valid only when the period or frequency of oscillation is gravity-determined such as in a pendulum. Obviously most biological periods or frequencies are not gravity-determined (pendulum-like). Some may be determined by other principles such as that of the relaxation oscillator. In such cases it is the electric property that determine the period, not the gravity. Thus the gravity-based "law" has limited significance in biology.

The theory of biological similarity is based on an ad hoc assumption $\tau = \lambda$. The concept of operational time seems just an empirical correction. It is difficult to find further significance of the theory besides being an empirical correlation.

GÜNTHER:

I would like to emphazise that for historical reasons NEWTON's theory of mechanical similarity represents the origin of all subsequent theories of biological similarity and therefore should not be eliminated on logical grounds. Moreover, in biology these theories of similarity have only a statistical meaning, since they pertain to the so-called "mixed regimes", where the mechanical similarity criterion ($\tau = \lambda^{1/2}$) represents one extreme of the Gaussian distribution and the other extreme is represented by the hydrodynamical similarity ($\tau = \lambda^2$); the mean value is equivalent to LAMBERT & TEISSIER's (1927) a priori assumption that $\tau = \lambda$. Finally, the mechanical similarity rules are valid in biology when gravity-determined phenomena are studied, as for instance the act of walking, which has been compared by D'ARCY THOMPSON in this book "On Growth and Form" (Vol. 1, page 39) with a physical pendulum. Recently, I found a physical basis for LAMBERT & TEISSIER's a priori postulate ($\tau = \lambda$), when relaxation oscillators (hydraulic and electrical models) of different size were studied, whose length ratios ($L_1/L_2 = \lambda$) were linearly correlated with the time ratios ($T_1/T_2 = \tau$). In conclusion, the a priori assumption ($\tau = \lambda$) has now a physical meaning, namely that in biology the chronological processes obey the rules of relaxation oscillations (paper in preparation). - The objection, that the "operational time" is only of empirical nature is correct, since its only function is to improve the predictive value of the theories of biological similarities. Nevertheless, it is likely that in the future this ad hoc assumption could eventually be associated with the "cellular" structure of all organism, a condition which is unknown in physical similarity theories.

SZEKELY:

Dimensional analysis must be applied within a total methodological circuit with closure for control. It is a somewhat <u>local</u> application of a technique within the <u>total</u> methodological circuit. The coordinative relation at the basis of dimensional analysis corresponds to LOCKER's request for relational basis within a more powerful language for biology. It seems to me that the essential point in the development of a Theoretical Biology is that of the design of a powerful code-language with relational basis (including at least one heterogeneous coordinative relation), a language which has been designed in accordance with the requirement of the methodological circuit for closure, control, evaluation and probable correction by feed-back. - I cannot find an answer to the following question, related to GÜNTHER's paper: If within the same living organism several organs have their own biological time - which of them is the dominating one and what equalizes the phase-differences - so that there can be one kind of governing biological time for the total organism? This is independent of the size of the body. This seems to be a system-theoretical problem, exceeding probably the present range of applicability of the classical Dimensional Analysis.

GÜNTHER:

As mentioned in our paper and illustrated by Fig. 2., within the same model of organism (mammals), heart and respiratory rates have in common the same allometric exponent of body weight (around -0.27), but the ratio between the corresponding parameters (a) for both functions is 1335/354 = 4.0, namely that one respiratory cycles has the equivalent duration as four heart beats. - There is no permanent command of one of these pace-makers on the others; rather a "Relative Coordination", in the sense of von HOLST, exists between the different relaxation oscillators (respiratory and cardiac, for instance). - Concerning the equalization of eventual phase differences among the relaxation oscillators of systems, organs, or cells, it is likely that the general metabolism per unit mass (see Fig. 2.) may help to synchronize possible phase shifts among various pacemakers, which are subordinated to neuro-humoral control mechanisms trying to preserve the organism as a whole, in the sense of Claude BERNARD's "ensemble harmonique".

LOCKER:
If we start with the assumption that a timeless order is the (formal) prerequisite for temporal processes, i.e. those occurring within time, then we make a distinction similar (or analogous) to that existing between being and becoming and, at another conceptual level, between relational and metric. Within the realm of processes that are time-variant or time-dependent the imposed metric determines the time-scale. However, several time forms can be delimited which are not determined by a metric but rather by relations. In this respect also a distinction between local time and global time seems to be justified. Whereas the former refers to the organism itself, being deeply rooted in the organism's organization (e.g. BACKMAN's time or your "operational" time) the latter refers to the total biosphere. Of course, both time forms must not be understood in terms of a physical so-called objective time and by global time in particular is not meant the elapsed time life required for its evolution during geological periods. What is meant is time referring adequately to its own subject. In conceiving the timeless (or supra-temporal) order as constitutive the time forms which are related to their subjects come forward as adaptive or, in other words, operative. - For COPE's law, which from a systems-theoretical view expresses the system's tendency towards expansion or extension of its space filling it appears not easy to decide whether the behavior according to it pertains to an operative or a constitutive time form. One might argue that the aspiration to space occupation is necessarily linked with the inverse tendency to reduce the system's complexity. WILLISTON's law shows this link of these both tendencies very clearly. Accordingly, an increase of complexity could be brought about by means of reducing the space filling. The occurrence of form (or size) exaggeration which is described by COPE's law could be interpreted as a preponderance of local time over global time with corresponding space seizure.

GÜNTHER:
If we assume that relaxation oscillators are basically related with all chronobiological processes, including the activity of pacemakers in different organs and of circadian endogenous clocks (WEVER), it seems that the imposed metric, studied on hydraulic and electrical models, determines the corresponding time scale. - We agree that time forms are created in agreement with the own subject, and a close correlation should be expected between local times - of organs for instance - and the global time (organism), as well as between the latter (local) and the biosphere (global). These relationship should be interpreted as of adaptative nature and not as fixed ones, since they can change according to the circumstances. - With regard to COPE's law, the space filling tendency of organism, during the evolutionary processes, should yield certain selective advantages, since any increment of body mass might be associated with:

a) an greater depot capacity for water, minerals and metabolites;
b) a reduction of the metabolic rate per unit mass, namely the "metabolic reduction" of LEHMANN;
c) a lesser "surface/volume" ratio, which is particularly important in heat-transfer processes in homeothermic animals;
d) an increase of the duration of all relaxation periods and consequently of the biological events associated with each cycle;
e) a greater life expectancy, since according to RUBNER's studies, there is a close correlation between the duration of life and the metabolic rate of homeotherms of different size; and
f) mechanical and physical advantages, which are associated with a greater body mass.

Timeless Order
E. W. Bastin

Abstract

Reasons are given for thinking that the kinds of temporal relationship considered in biological organization are too stereotyped. They derive their apparent necessity from our familiarity with the world of classical physics whereas new possibilities can be seen to be equally plausible a priori in the light of the - as yet undigested - changes in physics during this century. An outline model is given, which might exploit some of these possibilities - at least to the extent of illustrating the kind of freedom which exists. The model is sequential - taking the form of a simplified computer program.

1. Introduction

We have lived in the presence of a revolution in physics of enormous magnitude whose consequences have not been assimilated by the physicists themselves except in certain very limited directions. The revolution centrally concerns the status of time. The finiteness of the velocity of light accepted as a fundamental condition on observation - is the physicists' recipe for handling a situation of great potential conceptual novelty by establishing certain bounds + on space, time and velocity whithin which the conditions can still be thought of as "classical" and beyond which special and novel methods have to be sought. The trouble about this way of working is that the bounds have to appear as spatiotemporal - or at any rate as dynamical - whereas they are really of a logically different kind.

To exhibit the conceptual novelty clearly one should stress the limitations imposed on causal relationship of any sort by special relativity. The light cone (a construct involving a conventional space-time manifold and a limiting signal velocity) imposes a classification of events in respect of their having - or not having - causal relationship with an observer at the origin. Those in the light-cone may have them, those outside may not. The question then is, which to take as logically primitive, the geometry of the light cone or the classification of events. I plump for the classification of events, and regard the more conventional approach of taking the geometry as primitive as just a way to reconcile a conventional space-time picture with the conceptual novelty of the causal scheme, with as little modification as possible. One sweeps the conceptual novelty as far as possible under the carpet.

In this paper I will be concerned with the question of how it would affect our view of a possible science of mathematical biology to be explicit about the enlarged view of time-relationships that the changes in physics properly demand.

The "under-the-carpet" technique is not the only course open to us even within physics, though the current physics might lead one to think it was. Indeed a change within physics on a very big scale is now inevitable, chiefly because the "under-the-carpet" method can never deal with the

---

+ The "bounds", in this case, will not be defined by "relative velocity = c", but by "relative velocity is comparable with c" (say, relative velocity $\not> c/100$).

quantum problem. I have to make this assertion categorically, here, whereas it is of course contentious; however, that my assertion is now a recognized point of view is illustrated by - for example - many of the essays in "Quantum Theory and Beyond" (4).

Resistance to this change in physics can be traced to the implicit assumption that it is necessary to consider space as a differentiable manifold. It has been assumed that if A, B, and C are points, then there is always an intuitive meaning to be attached (operationally speaking) to the assertion that C lies between A and B. This assumption is the essential requirement for the existence of a differentiable manifold. Now we have seen that special relativity demands the existence of sets of points for some of whose members the assumption is not true, and if it is ever not true we are free to consider it as a logical possibility that it is not true more generally. If it breaks down locally, in particular, then one will no longer have an automatic ability to take a simply ordered sequence of events to define time. Then, it will be convenient to say that <u>classical temporality</u> breaks down. One should consider whether living organisms are structures which make use of this greater freedom.

We need not unravel the reasons why the differential manifold, and hence classical temporality, seemed so natural as to have the force almost of a logical necessity any further than to point out that it goes along with our sensory awareness of space through our manipulation with bits of the physical world. However, in high energy physics, where the assumption corresponds to no operationally practical situation at all, it is at present under scrutiny. CHEW (6) has discussed this, and more recently (7) has questioned the validity of the <u>completeness postulate</u> for HILBERT space on these grounds. In general one must conclude that because it is extremely important for the operational situation in a science to be correctly represented in its mathematics, once that mathematics has had to be abandoned for reasons of its operational inadequacy on one field there is a strong case for looking into its adequacy in the rest.

2. A General Alternative to the Differentiable Manifold

Instead of the classical continuum, with its appeal to intuition to describe the local environment of the observer, my own approach is to substitute the constructive space. By "constructive space" I mean a space where a rule is given for constructing new points as they are required to specify our knowledge as we come by it as a result of experiment or observation. The rule will in general depend upon the stage of development of the space that exists at the time in question. Among the thinkers who have contributed to the concept of a constructive space are WHITEHEAD (14) who developed ideas of spatial and temporal relationship in terms of ordering principles, and so explicitly posed the problem of replacing physical intuition by logical construction, and BROUWER (5) who considered "fans"-mathematical structures in which generating rules depended upon the epoch, or current state of development, of the mathematical structure. A general case can be made for thinking that the constructive space is the right general alternative to the differentiable manifold, for on general grounds quantum theory has shown that the interaction (or "observation") process through which we get information about the physical world must itself be incorporated in any adequate theory, and in fact the sequential character of the perception of the observer is made mathematically explicit in the constructive space. The more specific geometrical construction of special relativity with events outside the light-cone of an event at the origin, has the result that the indefinitely extended space existing independently of observation, which is retained in relativity, is really no more than a façon de parler.

Properly, one should consider that the universe only exists in any sense at all in relation to a given origin in terms of which observations can be posited. In this sense the sequential character of the constructive space has its implicit parallel at the base of relativity also. The new logical situation exhibited in common by quantum theory and by relativity is rendered explicit in the constructive space. The evolution of a constructive space can be traced back progressively to ever simpler configurations until we get back to the first steps, which - almost as a linguistic matter - we must attribute to a <u>constructive centre or "observer"</u> (not necessarily a personal observer). It is from this argument that the germ of the concept of <u>organism</u> appears in a constructive space.

Experimental phenomena are known which seem to demand a departure from "normal temporality" (a world in which every event can be mapped unambiguously into a simply ordered set of events which is chosen to represent "time"). Such are the remarkable T.P.C. equivalences in high energy physics which demand a fundamentally new concept of time, and "time reversal" is usually thought to be the only possibility of change. In a quite different field the many well-attested cases of precognition seem to demand that we abandon normal temporality and consider a world with flexibility in assigning temporal order "locally" (as contrasted with the freedom of this sort allowed in special relativity non-locally). Nevertheless, however important these experimental facts may be in showing that we have got to develop concepts capable of handling something more sophisticated than normal temporality, it is doubtful whether we should expect to advance by concentrating our attention on these extreme cases. They are the cases which will not fit into the old picture at all, whereas what we should probably study is something capable of being thought about both in the old terms and in the new.

The systems to which I look to provide guidelines into the intermediate region are ones that are generally admitted to be <u>evolutionary</u> in the sense that we need to consider a gradual past development in order to understand them as they now are. If we assume classical temporality, then we usually also suppose that this evolutionary development is sufficiently separate from the structure:

1. to be there to investigate whenever we need to, and
2. to be capable of being supplemented by environmental information (other members of the same species and traces of ancestors etc.)

The perplexity (and of course the fascination) of getting away from classical temporality is that whatever we need of evolutionary development for understanding structure must be explicity present in the structure, for we no longer have a "time-like-an-ever-rolling-stream" sort of past to appeal to. If at this point the critic objects that evolutionary - for example biological - structures exist and work without help from the past of their species, then we have to reply that he is imposing a very concrete idea of what the structure at a given time actually is, and that leaving aside the question whether he really has enough knowledge to presuppose this, we need a much greater flexibility. In this paper I shall consider a class of mathematical models which are designed to supply this flexibility. They will be called event-memory systems.

3. Event-Memory-Systems

Event-memory systems are hierarchical in a specific sense - namely that an event constructed out of a memory and a physical event (and itself given a memory record) is itself an event.

I shall call such a hierarchy an <u>event-memory hierarchy</u> to distinguish it from the classificatory hierarchy such as was considered in a recent symposium and published as Hierarchical Structures (15). Event-memory hierarchies were described by myself (2) and again there is a past history to the concept in the thinking of BROUWER (5). GÜNTHER (9) has worked on related ideas, though more in the context of logics. One should cite also the mathematics of categories in which a set is extended to include mappings onto the set.

I made some general suggestions about how to reconcile such a structure with the more usual <u>level</u>-hierarchy of the sort described in "Hierarchical Structures". The suggestions amounted to using time constants as the criterion for <u>level</u> membership. This treatment arose from contact with related ideas of some biologists at the Bellagio conference of 1968 +, who were interested in characteristic periodicities in living organisms (IBERALL (10), GOODWIN (8)), and was connected with the underlying idea of the memory hierarchy through the idea of <u>discrimination</u> as I now summarize.

In a dynamic p-simplical-complex representing the state of a system taken together with a specification of its history (ATKIN and BASTIN (1)) it is possible to use <u>every possible</u> complex up to p to represent a configuration of a system outside the original one. This is a logical limit which we describe as <u>maximal discrimination</u>. It is not possible to define any memory in a maximally discriminated set since any memory specification would require the use of a new complex as a memory point, and all have been used. Such a system can trivially be used to define a simply ordered time since there is no possibility of conflict, and has been used (1), (3), (14) to attack the physical problem of the origin of quantization in a numerical manner. The fact that this use has been made is important for my present application, since it provides me with something fairly definite to try to generalize, and hence gives a guideline in tackling a particularly difficult mathematical and conceptual problem.

4. Computing Models as Evolutionary Systems

The simplified picture with classical temporality corresponds with the assumptions of Darwinian evolution. In the terminology of § 3 this means that a fixed time sequence is defined by distinguishing individuals in a species at different stages of evolution. Having done this, a class of individuals that are "adequately similar" to act as parents in a reproductive process is defined. In our normal way of thinking, these exist contemporaneously, and no definition of adequate similarity has to be provided, but in a generalized evolutionary system no such appeal to classical temporality is possible. The species is defined as the whole potentiality for development of an individual except in so far as restrictions of a sort characteristic of the required species are imposed. The whole set of 1,2 ... , p-1 complexes existing and available through a memory system, or any subset of them, can be concatenated to generate a new p-complex. Thus in principle any number of existing individuals can be concatenated, providing the result is only a p-complex, in a generalization of the reproductive process beyond that in which two individuals of adequate "similarity" act as parents.

Model 1., which follows, is a set of instructions designed to be programmed in the high level computing language TRAC due to MOOERS and DEUTSCH (12). The most important functions in this language for the present purpose have the following names and simplified descriptions:

1. String: An ordered set of characters or ciphers.
2. Store: A set of ordered pairs, each consisting of one string with one name.
3. Call: The operation of drawing a string from store by specifying its name. The string remains in store.
4. Define: The operation of attaching a new string to a given name (and therefore of placing the string in store). A string defined with a given name replaces any string already in store of that name.
5. Equals: A test for identity of two strings.

It can be seen in a general way that these operations are suitable for programming Model 1. Even more importantly, TRAC implementation makes it natural to think in a way appropriate to an event-memory hierarchy. The programme can be written using a name for any piece of programme

---

+ In an editorial comment (15) WADDINGTON specified a concern with time constants as one unifying theme that emerged at the conference.

we have not yet decided how to do - leaving it to be evaluated later - possibly in a way depending on a later stage of the programme. Without this facility, sequential operation without beginning from the beginning each time would be impossible.

A TRAC programme to represent simple Mendelian inheritance has been written by McKINNON-WOOD (14) but is too long to be included in this paper. Subsequent work will consist in investigation of the stages by which Model I. has to be specialized to get something comparable with McKINNON-WOOD's model.

Model I.
---

1. There exists a set $I = I$ (A, B ... to h terms) of binary strings (strings consisting entirely of the binary units 0, 1) of equal length l, in store at a given time.

2. There exists a set of names which may be assigned to the strings in (1). These are equal strings a, b, ... to h terms of length n, it being convenient to give the name a to string A.

3. A generating process α will be defined (in this and the following rules) for constructing members of I. If P and Q (P and Q in I) are called, then a new string R is constructed such that for all P, Q there exists a unique R, and such that: -

4. R is of the same length as P and Q, and such that:-

5. If and only if P has the same binary unit in each given place as Q has in that place (i.e. P and Q are identical) R is the zero string (consisting of zeroes in each place).

6. No preference exists for particular places in either string.

7. The operation α acts elementwise. That is to say the pth element in one string is taken together with the pth in the other to construct the pth in the resulting string.
(From the foregoing rules the generating function α can be narrowed down to symmetric difference).

8. A sequential process is started by calling non zero strings P and Q successively to generate R. We write $(PQ) \longrightarrow R$.

9. A second generating process β is applied after the process α and is determined by the names q , r in a manner to be specified (rule 12 et seq.).

10. $\beta_{\nu\mu}$ is treated as the name of a new object described by a new string, and the set of such objects is said to be at a new level.

11. The relations defined between two levels can be repeated to define adjacent levels in the direction of greater complexity. There may also exist levels less complex than the original.

12. The generating process β q, r (rule 9) reduces to β q in the case when no name r for R exists, and in this case the string q replaces Q (which - having no name -makes no difference to the total construction).

13. When r exists, a string of length l^2 is constructed as a function of q, r. This is treated as the name of a new string at the new level. This function is chosen in such a way that the name of each object at the preceding level (with an upper bound of $2^l - 1$) in its relation to the object, is preserved.

Comments on the Rules

1. We imagine we are breaking into a hierarchical construction process at some stage.

2. The relation of the length of name-strings to the length of object strings cannot be defined in general since it depends recursively upon later rules (see rule 12). In certain special cases it is possible to pursue the hierarchy to simpler levels and reach a bottom level at which the relation between name-string length and object-string length can be fixed. This then dictates that relationship at other levels.

5. The combination of this rule with rule 12 gives a unique status to the zero-string. This uniqueness is a logical property of memory-event hierarchies in general. The zero-string is used as a name in the construction of higher levels (generating process β), but is treated as a non-existent element at the existing level (generating process α). This logical property was formalized by AMSON (see (7)) in the context of set theory by distinguishing closure under a new operation called "discrimination" from closure of the set. Later (1) ATKIN formalized it in the context of a simplicial complex by extending a set S of discriminated points in a totally disconnected complex by including the cone point $\emptyset$, and identifying the discrimination of a given point A in the initial set with that of the pairs ($\emptyset$, A) and (A, $\emptyset$) in the extended set. In both cases these procedures make no sense in a static mathematical structure.

13. Preceding rules are specific and yet (given the general type of structure we are designing) do not restrict generality. Rule 13 is a non-specific condition allowing a wide range of possible specific rules. At this stage we abandon the attempt to deduce specific forms.

References

1. ATKIN, R.H., BASTIN, E.W.: Int. J. Theor. Phys. $\underline{3}$ (6), 449 (1970).
2. BASTIN, E.W.: Stud. Philos. Gandensia $\underline{4}$, 77 (1966).
3. BASTIN, E.W.: In: Contemp. Physics Vol. 2, p. 451, Vienna: Int. Atomic Energy Agency, 1969.
4. BASTIN, E.W. (Ed.): "Quantum Theory and Beyond", Cambridge 1971.
5. BROUWER, L.E.J.: Can. J. Math. $\underline{6}$, 1 (1954).
6. CHEW, G.F.: "Closure, Locality and the Bootstrap", Laurence Rad. Lab. Reprint 1967.
7. CHEW, G.F.: In: E.W. BASTIN (1. c. (4)), p. 141.
8. GOODWIN, B.C.: In: C.H. WADDINGTON (Ed.): "Towards a Theoretical Biology" Vol. 2, p. 140 (1970).
9. GÜNTHER, G.: Akten des XIV. Intern. Philos. Kongresses, Vol. 3 , p. 37 , Wien, 1968/1969.
10. IBERALL, I.: In: C.H. WADDINGTON (Ed.): "Towards a Theoretical Biology", Vol. 2, p. 166 (1970).
11. McKINNON-WOOD, R.A.: "A TRAC Model for Mendelian Inheritance", Cambr. Language Res. Unit. Workpaper, 1971.
12. MOOERS, C.N.: TRAC, a Text-Handling Language, paper presented at the 20th Natl. Conf. A.C.M., Cleveland, Ohio, Aug. 1965.
13. WADDINGTON, C.H. (Ed.): "Towards an Theoretical Biology", Edinb. Univ. Press, 1970, p. 265.
14. WHITEHEAD, A.N.: "Mathematical Concepts of the Material World", Phil. Trans. Roy. Soc. 1906.
15. WHYTE, L., WILSON & WILSON: "Hierarchical Structures", Elsevier, New York, 1969.

DISCUSSION

BREMERMANN:
This paper is nothing but a piece of philosophical rumbling about the foundations of physics and it has nothing to do with the theme of this book. I doubt that it would be publishable in any physics journal.

BASTIN:
Dr. BREMERMANN accuses me of having produced "nothing but a piece of philosophical rumbling" (and whether or not he meant to write "rambling", the suggestion of noises in the tummy with mixed overtones of "rambling" and "bumbling" is too good to waste). If there were a lot of people experimenting in the sort of way that I am, then I would agree with BREMERMANN that my effort was too remote from experimental detail to make it worth including in this book. As there are not, I think it was worth trying to make the case for "philosophical" models in this context by producing the bare bones of one. - I think it quite obvious that the extrapolation of the known methods of physics and chemistry is never going to provide principles of organization adequate to the understanding of living things. I think this not because these principles have to be of some different, vital, sort, but because to get to them you have to go back to a more general class of relationships between entities of which the physical and chemical are specialized cases - and specialized in the wrong way. Naturally I do not expect anyone to be influenced just because I happen to think that, but equally I do not expect my view to be rejected on a priori grounds either, and such unreasoning rejection often takes the form of castigating whatever is speculative as "philosophical". This use of the word "philosophical" carries the strong suggestion of representing the opposite of whatever is physically real. A further step in the same train of thought has been taken by those who identify "reality" with that which can be deduced from current physics and chemistry. It is ironic that people (and I do not assume BREMERMANN is in this class) who have taken this last step should glory in their lack of philosophical sophistication when they stand so urgently in need of it. - In a better and more educated sense of "philosophical" it requires considerable philosophical expertize to handle speculative models safely, and I see little point in this book if in it we are not prepared, and able, to be philosophical in that sense. -
In questioning the relevance of my note to this book, BREMERMANN might reasonably expect me to make remarks about how a highly generalized memory of the sort I have considered comes to be incorporated in a physical organism and associated with neural tissue. However, I plead shortage of space to justify my having restricted myself to trying to establish the reasonableness of my approach in very general terms (a difficult enough enterprise in itself). Some of LOCKER's remarks relate to this gap and he gives me some encouragement at a metaphysical level.

ANDREW:
Most people find it difficult to think about any physical or biological system without tacitly assuming that time is like an ever-rolling stream. Clearly this is in pretty good accordance with experience, as are also the assumptions of Newtonian mechanics. There is the interesting difference, however, that whereas departures from Newtonian mechanics are almost impossible to demonstrate without the elaborate equipment of high-energy physics, evidence for departures from classical temporality are also provided by the phenomena studied in Psychical Research. The departures in these phenomena may be quite large; that is to say, the time displacements can be days, months or years. - I am wondering whether BASTIN would like to hazard any guesses about future developments. Will time travel at will become a reality and not just science fiction? Might a way be found of travelling to other solar systems and other galaxies without spending hundreds of thousands of years on the way? These may seem rather frivolous questions to ask about a topic having deep implications for modern physics, but I would like to hear BASTIN's views.

BASTIN:
Dr. ANDREW is certainly trying to entice me in the opposite direction from that in which BREMERMANN would have anyone go. I do agree with all his remarks on time and our assumptions about it, but as to speculating further I think I should limit myself to what my model suggests (as if that isn't already bad enough). - I was imagining in the first place a completely unrestricted field of relationships between entities of some primitive kind. I was then considering whether some sort of memory structure which could be characteristic of a living organism (not necessarily a <u>conscious</u> memory structure) could entail some degree of limitation on the field of primitive relationships in such a way as to make it possible to define a time with some degree of universality or, at any rate, of consistency. According to this suggestion, therefore, you have space-time perception linked with organic structure (in a very wide sense of organic structure which includes memory structure) and if you want to transcend spatial and temporal limitations so as, for example, to have time travel, then you would have to renounce your whole constitution as a physical or living organism, and I'm not sure whether one could properly call that "travel". (It does seem however to bear some relation to what happens under the influence of hallucinogenic drugs, or even in dreams sometimes) To explore such question further would need far more sophisticated models than my present one.

LOCKER:
Your paper contains some very interesting items. It offers an important break-through to a conseptual innovation by linking up evolution with constructivity. The guiding star of your exposition obviously consists in the irrefutable reference of constructivity to a (subject- like?) centre. You make the range wherein evolution is possible extend between something that is given (i.e. in your consideration: normal temporality) to anything other that is so flexible as to allow the emergence of every "desired" local temporal order. This idea vividly reminds me of the relation between actual and potential, the latter being the spring for the perpetual evolution (i.e. the becoming) of the being. I understand, it is to be hoped correctly, your conception of structure in the following way: Any structure can be conceived of (and thus be regarded as concretized) at any arbitrarily given time point, but whenever it is viewed so-to-speak from its interior as evolving within (objective) time it is able to create its pertinent time as an operative time. Clearly, in close parallelism with a spatially formed (and imaginable) hierarchy, another kind of hierarchy, namely a temporal, must be postulated. However, the role you are ascribing to memory, in connection with a physical event, in enabling another event to arise, is not entirely translucent to me. Since this kind of hierarchy ought to be (in principle) infinitely continuable, the newly arisen event must also again carry along with itself the past (i.e. the memory plus the physical event) and so forth, notwithstanding the possible meaning of these terms; otherwise a true hierarchy uniting at each level (or moment) all the subordinated (or past) could not be built-up. It is reasonable to call upon time scales as criteria for level membership (in the spatial hierarchy) (and conversely for space measures as criteria for moment membership in the temporal hierarchy), but where are these criteria steming from? How do you distinguish between complex, and configuration and system? One could easily imagine the existence of a sequence of specifications, starting with a complex usable to represent a configuration, which in turn representatively refers to a system. The term discrimination seems to indicate that all of the possible configurations have been employed as representatives of the system. Another expression for maximal unfolding? I understand memory point with respect to memory specification in some analogy to an Archimedian point which as an undispensable footpoint is the prerequisite for any construction. - Another important statement made in your paper is the contradistinction between differentiable manifold and constructive space. Here you touch upon the foundations of mathematics which today - at least in a general way - have definitely achieved a clarification in unique favour of constructionalism (intuitionism or operationalism). A similar situation characterizes ontology, as recognizable in the search for the ontological basis of General Systems Theory shown in a paper devoted to L. v. BERTALANFFY, and therefore sheds light onto the problem of biogenesis, too.

BASTIN:

I should first try to clarify the relationship between what LOCKER calls my "spatial" and "temporal" hierarchies. The spatial hierarchy (that is to say the one that refers to the spatial concepts familiar from classical physics), is obtained as a specially simple case from the general construction of my paper. The simplifying feature of the spatial hierarchy is that in it <u>all</u> the possibilities of construction are <u>always</u> realized. I have referred to this as a situation of <u>maximal discrimination</u>. This case is simple because no provision has to be made for specifying the choices at each point of the construction process, since one will continue till all possibilities have been followed. In this special case we have found that the numbers of discriminable entities at the successive levels of the hierarchy give a good order of magnitude agreement with the strengths of the basic fields of physics – strong, electromagnetic, gravitational – and that the construction then terminates. This correspondence with the remarkable field structure of physics may be strong enough to justify thinking of the maximal discrimination hierarchy as providing a physical (spatial) background. – Then we come to the more interesting general cases which LOCKER calls "temporal". Here we have to postulate a "store" or "memory", for it is no longer the case that we are not interested in the decisions. Hence each decision must be made either by appealing to a rule which lies outside the system which we therefore cannot know anything about (and this includes randomization) or else the rules must be part of the system. We are naturally interested in the latter case, and in that case the rules on which decisions are made will in general depend on past configuration of the system. Information on this matter constitutes the memory. Thus I do not see any <u>logical</u> difficulty in postulating a memory. (Or course there are scientific difficulties – for example in postulating a memory prior to constructing physical structures in which it can be housed – but at the moment I am only constructing a metaphysical model and am more than happy if I can get such a model logically possible so that it can stand as an alternative to the metaphysical model which is currently presupposed) – Now, however, we get to the real point which LOCKER is raising. There has – he argues – to be another "time" which we could call "epochal time" or "epoch" through which the whole system (as distinct from what he calls the Archimedian point) progresses. I fully recognize the need for this concept, and I conjecture that it can only be defined in the case of hierarchies with a considerable degree of specialized structure. Unfortunately I have little idea of the mathematical properties which might specify such structure. Possibly we human beings – as highly organized systems – get practical acquaintance with this distinction between time as the successive moves of the Archimedian point and time as epoch, in the distinction that we can make experimentally between our succession of conscious states on the one hand and our more general sense of progress through life on the other.

The Ontogenesis of Purposive Activity
A. M. Andrew

Abstract
People and animals have a remarkable ability to develop skills from experience. Ways of studying this are discussed, with particular reference to its imitation in artifacts, which may sometimes be regarded as embodying hypotheses about the working of the nervous system. It is suggested that thought processes depending on language are less important than is often suggested. Possible constituent elements and fundamental operations of a nervous system are briefly reviewed. Some principles of very general applicability are treated, particularly the reduction of signal redundancy and the use of a principle termed "significance feedback" to control adaptive changes in a self-organizing system.

People and animals are able to survive in environments containing threats such as predators and precipices, and in which skilful activity is needed to win food and other essentials. This capability depends on complex information processing by their nervous systems, which must store a large amount of information to specify the nature of the processing carried out. Some of the stored information is genetically determined and some is ontogenetic, or resulting from previous experience of the individual. In experimental psychology the attempt is often made to determine the relative extents to which a skill is inherited and learned. The answer is seldom clear-cut, since the two kinds of information are not held in separate watertight compartments. Sometimes an indication of their relative importance can be obtained, however, as when animals are tested for a skill after being reared in an impoverished environment. An example of such a study has been reported (7) which will be discussed later.

Some of the information acquired by experience takes the form of memories of discrete events, at least in the case of human beings (see (11)). The genetically-determined information does not contain recollections of particular events; it is rather in the nature of skills, or policies needed to determine outputs as functions of input signals. Much of the ontogenetic information is also of the "skill" or "policy" kind, even though it results from a succession of discrete events in the individual's experience. For example, a driver of a car approaching a bend does not normally begin there and then, to reflect on the previous occasions on which he approached similar bends, and to process a mass of stored information on how he reacted and what was the outcome each time. Instead he uses information of the "skill" type derived from the past events.

The ability of people and animals to acquire skills by experience is well beyond anything so far achieved in artifacts. It is only necessary to refer to the fact that people learn to read cursive handwriting of poor quality, and people and animals to ride monocycles, to illustrate this. There is therefore much interest in trying to devise "learning machines" with some of the same capabilities. Such machines could have direct practical utility and some aspects of their operation might constitute useful hypotheses about the operation of nervous systems.

Self-organizing Systems

Attempts to produce something which might reasonable be called intelligence in artifacts fall into two classes. The kind of work generally described under the heading of "<u>Artificial Intelligence</u>" makes full use of the facilities of a digital computer and the tasks performed include the proving of theorems and the playing of chess and other games. The serial nature of the operation of a computer lends itself to the simulation of those aspects of thought which are accessible to introspection, and the achievements of these studies are impressive. Many of the most impressive programs do not embody the means of learning by experience, though the famous program of SAMUEL (18) for playing checkers (draughts) is one which does.

The other approach is the study of self-organizing systems. Work under this heading is on the assumption that there are principles of self-organization sufficiently general to be embodied throughout a complex network and effective in modifying it in a goal-directed way. There seems to be little doubt from neurophysiological studies that some such principle or principles do operate. Among the questions which remain unanswered are those of whether there is one principle or many, and how locally within the network it (or they) operate.

The success of the "Artificial Intelligence" approach in the context of certain types of problem can be partly attributed to the fact that there is generally no such thing as a "<u>local operation</u>" within the system. Any item of information available to the system can, in principle, be brought into interaction with any other. Similary, when a person thinks about such problems, he can consider any part of a large store of information in conjunction with any other part; a holistic approach is possible which seems unlikely to be achieved in a self-organizing system.

A person tackling problems in theorem-proving or game-playing makes use of language in his thinking. In fact, the problem can only be specified by the use of language. It also seems reasonable to suppose that it is the use of language which makes possible the holistic approach referred to above. The writer would like to suggest, however, that the importance of language in the study of information processing in the nervous system is not so great as is sometimes suggested. The acquisition of many skills, both perceptual and motor, has little if any verbal content. This is emphasized by the fact that many of them can be learned by animals. The question whether the types of cognitive development studied by PIAGET depend on language is discussed by SINCLAIR-de-ZWART (19), who decides they probably do not, though the issue is far from clear-cut.

Even in the context of the "higher" skills such as chess-playing and theorem-proving it is debatable whether that part of the thinking which is expressed in language is the whole story. DREYFUS (9) argues very convincingly that it cannot be. His argument is a reversal of what was stated above about the holistic nature of the processing permitted by a computer program or a person reasoning linguistically. He argues that in performing the "higher" tasks it is frequently necessary to be sensitive to a great many items of input data, as well as tentative, incompletely-formalized developments of them. The nervous system takes them into account by some form of parallel processing, but a sequential process like the operation of a digital computer or a strictly linguistic reasoning process cannot do so in any reasonably short interval of time. If this is so, human thought must depend very much on non-linguistic processes, even if the outcomes of these are promptly rationalised in verbal form.

The initial acquisition of a language is an example of a complex process occurring in the nervous system without, in the initial stages at least, any dependence on the use of language. CHOMSKY has suggested that tacit knowledge of the general structure of language is inherited, but even so it is a formidable task to learn the substantial part which must clearly be learned since it varies from language to language. The task involves much more than simple associative learning. DREYFUS illustrates this quoting WITTGENSTEIN, by pointing out that if we show a child a table and say "brown", there is no simple way in which the child can be sure whether "brown" refers to the colour, the size or shape, or is a name for the type of object (as "table" is) or a proper name (like "Mr. Brown"). Nevertheless, children do learn to use adjectives, common nouns, proper nouns, etc.

These considerations show there is no reason to reject the self-organizing system approach on the grounds that it is non-linguistic. It could, of course, be argued that it is in any case supported by the neurophysiological data and that this further justification in unnecessary.

What Sort of System?

There are, however, good reasons for abandoning the old idea that all that is necessary is to put together a sufficiently complex network of replicas of some kind of model neuron, perhaps with a mechanism for facilitating transmission through the most frequently-used synapses, and interesting properties must appear. Perhaps the idea has never seriously been put forward in exactly this way, but something of the sort has frequently been implied. The proposal of HEBB (12), as well as the systems considered by some of the pioneers of neural modelling such as BEURLE (5), were rather more sophisticated since they postulated patterns of sustained (reverberatory) activity within a cell assembly or network, coupled with synaptic or threshold facilitation which made frequently-occurring patterns more likely to occur again. BEURLE's system also allowed for feedback from the environment to dampen disadvantageous patterns and to encourage useful ones.

There is some evidence that networks might embody structure which would make it appropriate to regard something other than the simple threshold-element model neuron as the basic unit. It has been shown that in various cortical areas of the brains of different animals, there are ensembles of neighbouring cells within which activity is highly correlated from cell to cell (15). This would indicate redundant operation which could provide reliable response from unreliable components, as discussed by WINOGRAD and COWAN (24), and might make it appropriate to treat the <u>ensemble</u> as the basic unit in a self-organizing system. The extensive study of area 17 of the visual cortex by HUBEL and WIESEL (13) has revealed a great deal of structure which is presumably genetically-determined and this would suggest that the appropriate basic units would be sub-structures of this. (Recent work (7), however, shows that the structure is more susceptible to modification by individual experiences, at a rather fundamental level, than originally appeared. These workers reared kittens from birth with visual experience only of a pattern of horizontal stripes, and another group with experience of vertical stripes. In each group the cortical units as described by HUBEL and WIESEL were found to have orientations corresponding to the visual environment.)

Some very interesting possibilities arise if the system is allowed to have special pathways within it whose purpose is purely to guide the self-organization. These will be discussed in a later section.

Describing Brain Activity

The brain is so versatile that it is difficult to make any general statement describing its activity. In at least this respect it resembles a digital computer which has received the attention of an imaginative group of programmers. Different workers have chosen particular aspects of brain function for study, believing them to be fundamental. Some, for instance, have emphasized the specificity of the response, and the fact that a very small change in sensory input may produce a radical change in the output activity. Others have emphasized the continuous relationship which may exist between input and output, so that small changes in input lead to small changes in output. An example of the latter type of situation is seen if we consider a person running to catch a cricket ball he sees in the sky. In general, if he were to see the ball in a slightly different position, or moving with slightly different velocity, his pattern of response would be only slightly affected.

The ability to form concepts (6), (14) has some claim to be regarded as fundamental. The brain is required to interpret its environment as a finite automaton without memory, and to form a similar automaton inside itself. The internal automaton should then produce outputs agreeing with the indications of concept appearance coming from the "teaching input".

Another type of study, which has been undertaken by a group in Moscow State University headed by A. V. NAPALKOV, is in a way more general than the studies of concept formation since the brain is required to interpret its environment as a finite automaton which may have memory. An animal or subject is placed in a situation in which it is rewarded if it performs a sequence of responses, say a, b, c, d. Many variations are possible; for instance these actions may or may not be separately reinforced by signals from the environment, and the condition for a reward may be either the occurrence of these actions in strict sequence or their occurrence interspersed with others, as in a, p, b, q, c, r, d. In fact, a whole range of tasks can be presented, most readily described by drawing a flow diagram whose nodes correspond to possible states of the environment. The process of learning to obtain the reward can then be interpreted as the gradual discovery of the flow diagram by the animal or subject.

In the case where the subject's actions are separately reinforced, the tasks can alternatively be described in terms of chains of conditioned reflexes, and some algorithms by which the subject or animal might build up an appropriate chain were described earlier (8). The group is now studying brain algorithms in a still more general way, and is looking for a set which can be combined to account for a wide range of behavior. Algorithms to evaluate and modify other algorithms are to be included.

None of the particular aspects of brain function studied has been shown to be fundamental to all brain functioning. In the next section some ideas will be discussed which are thought to have very general relevance.

Significance Feedback and Reduction of Redundancy

It is assumed that any self-organization which occurs in a nervous system is goal-directed and consequently that the system is sensitive to some measure of a degree of goal-achievement resulting from its activity. Previous work on artifacts by ROSENBLATT (17), GABOR (10) and others (see (2)), has shown some ways in which the changes within the network can be determined by a feedback of such a measure. In the schemes of ROSENBLATT and GABOR, however, all the adaptive changes occur at locations which contribute in a very simple way to the output of the device. In a simple perceptron, as described by ROSENBLATT, the parameters which are adjusted are the "weights" associated with the pathways by which association units contribute to the final summation in a response unit. In a somewhat similar way, the parameters adjusted in a GABOR filter are coefficients determining the size of contribution to the final sum of the separate polynomial terms.

The restriction of the adjustments to a single layer imposes a serious limitation on the capabilities of the system, as has been shown conclusively for the simple perceptron (16). Their results depend on the very reasonable assumption that there is some limitation on the set of sensory elements which may be connected to a single association unit. If there is no such limitation, and no limitation on the total number of association units, a perceptron can, in principle, discriminate any equivalence class whatsoever, simply by letting one association unit correspond to every pattern in the class. In interesting tasks, however, the equivalence classes are so large that this is not practical.

Because of these restrictions it is interesting to consider systems in which the adaptive changes are not restricted to a single layer, especially since systems free from such restriction are more likely to be useful analogues of the nervous system. For the unrestricted systems it is difficult to find a suitable "training algorithm" to determine the adaptive changes.

One possible solution is given (in two variations) by STAFFORD (20). Each of his methods requires that changes be made in the net which are either temporary or permanent according to how they are found to affect the output of the net. This means that if the net is part of a system controlling, say, a group of muscles, it is necessary either that the net can be disconnected from the muscles while adjustments are made, or that the adjustments depend on activity in "shadow pathways" duplicating the main ones and used purely for testing possible changes. In either case it can be said that "dummy runs" are needed in order to arrive at appropriate adjustments. This seems rather cumbersome and implausible as a model of what happens in the nervous system, but should not be ruled out.

An alternative method (1) depends on special pathways in the net providing data for a continuous, automatic sensitivity analysis (21) throughout it. That is to say, at each point in the net a measure is available of the sensitivity of the output of the net to activity at the point. Such a measure can be used to determine the adjustment made at the point.

To see how such feedback could operate, and what sort of structure is needed to implement it, consider an element of the net whose inputs are continuous time-varying signals $x(t)$ and $y(t)$ and whose output $z(t)$ is the product of these (Fig. 1.). There would be a feedback signal $s(t)$ associ-

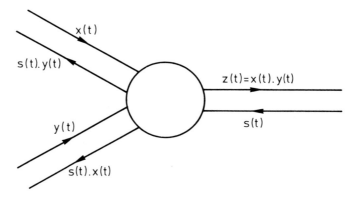

Fig. 1. Multiplication element with significance feedback

ated with the pathway transmitting $z(t)$, indicating the sensitivity to it of some output pathway. Then the feedback signal associated with the pathway transmitting $x(t)$ would be $s(t).y(t)$ and that associated with the pathway transmitting $y(t)$ would be $s(t).x(t)$. The element which performs the multiplication of the forward-going signals $x(t)$ and $y(t)$ must incorporate the means of performing the appropriate operations on the feedback signals, and all elements in the network must operate in this dual way.

An element performing multiplication has been considered as an example since the way in which such an element must operate on the feedback signals can be readily seen. Difficulties arise if the net contains elements which introduce time delays, or elements giving a strongly non-linear relationship between any one input and the output of the element.

The difficulty with non-linear elements arises because the sensitivity of the output to one of the inputs depends very much on the amplitude of the input signal. Model neurons of the McCULLOCH-PITTS type are threshold elements and therefore strongly non-linear and the difficulty arises very strongly in connection with them. Nevertheless, approximate ways of computing a sensitivity measure throughout a network of model neurons have been devised. Tests with a computer simulation of such a network have shown them to be effective in producing improvement from some states of the network, but at present there are non-optimal "trapping states" from which no further progress is made. The work is at an early stage and these difficulties can probably be overcome. At present a study is being made of the application of the method to networks of elements dealing with continuous signals, like the multiplication element mentioned above. Such networks have potential value as "learning filters", and their study may indicate some general principles also applicable to networks of threshold elements.

The term "significance feedback" has been used to denote the kind of feedback employed in the continuous sensitivity analysis discussed above. Although there is no direct physiological evidence for its occurrence in the nervous system, it seems the simplest way in which goal-directed self-organization in a complex network could occur. In social and industrial systems also, people forming part of the system are presumably strongly influenced by what they believe to be the effect of their activities on the behavior of the system as a whole, and hence by a form of feedback analogous to that postulated.

In the case of these systems it is often profitable to study and, where appropriate, to modify these feedback pathways. BARLOW (3), (4) and UTTLEY (22), (23) have also appreciated the difficulty of achieving goal-directed self-organization in a complex neural net sensitive only to a measure of overall goal-achievement. They have suggested that the changes within the net might be determined by goals which are evaluated more locally, particularly that of reducing the redundancy of the signals being processed.

Sensory input data is likely to be highly redundant, and the suggestion that the redundancy is reduced in early stages of processing agrees well with experimental data. The familiar observation that the response of all parts of the nervous system to abrupt changes in signal level is much greater than that to sustained inputs is an example of redundancy-reduction. The change may be abrupt in either time or space. A signal which does not indicate a change with time or in space is one which could have been inferred by extrapolation of other signals, and is redundant.

It is not necessarily obvious, however, what constitutes redundancy in input data. Signals appearing in biological systems, like those in artifacts, are normally influenced by processes which are irrelevant to the goals of the system; the term "noise" is used to designate the effects of these. It is impossible to decide, by examining only the signals, which parts of them are noise and which are significant. Hence redundancy, like other information-theoretic measures, cannot be estimated without reference to overall goals.

Examination of the signals without such reference may set a lower limit to redundancy. For example, if two parallel channels convey signals which are indistinguishable from each other, then clearly there is redundancy. If, however, there are differences between the two signals, the estimate of redundancy depends on whether the differences are ascribed to noise or regarded as significant.

The difficulties inherent in the use of redundancy-reduction as a locally-computed goal are illustrated by the fact that in the discussions of such possibilities it is not clear whether units should come to respond selectively to frequently-occurring patterns of activity in their input pathways, or to infrequently-occurring ones. From one point of view the former alternative is to be preferred, since these are the patterns to which appropriate response patterns have been evolved. On the other hand the infrequent patterns have greater information content provided their differences from frequently-occurring patterns are significant for the overall goals of the system.

If parts of a network are to undergo changes which will cause them to reduce redundancy in a useful way, therefore, they must either be sensitive to overall goals, or else must only be effective for certain kinds of redundancy. Redundancy-reduction is therefore likely to be most effective when operating in conjunction with some form of "significance feedback" as discussed earlier.

Conclusions

There is some very indirect evidence for "significance feedback" in the nervous system from the study of attention mechanisms. In cats exposed to a repetitive auditory stimulus, for instance, direct recording of electrical activity has shown that the signals are passed on by the cochlear nucleus with greater amplitude when the cat is free to attend to them than when it is distracted by an olfactory stimulus. Since the cochlear nucleus is well removed from the parts of the nervous system usually thought to be possible seats of consciousness or attention, this is presumably due to a feedback indicating the significance to the animal of the information in the auditory pathway. Although this is something which might merit the term "significance feedback", however, it is rather far removed from the form of it which could operate at the cellular level to modify synaptic strengths or other characteristics of single neurons.

At the time of writing, the author has put aside the attempt to improve the multi-layer adaptation of networks of neuron-like elements and is examining networks of continuous elements like the multiplication element of Fig. 1. The network of Fig. 2. has been simulated; each of the

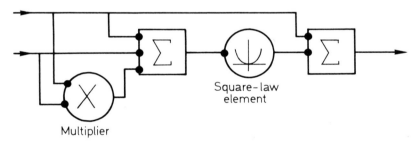

Fig. 2. Non-linear network used in study of automatic adjustment controlled by significance feedback. The black dots show points at which adjustment can take place

points at which a signal pathway enters an element of the net has a variable gain associated with it. The black dots in the figure show these points at which adjustment is possible. The output of the adjustable network is compared with that of a similar one having fixed settings for all the gains, and in a succession of trials with random inputs to the networks the variable gains are adjusted via "significance feedback" pathways so as to bring the two outputs into agreement. The adjustable network forms a non-linear "learning filter" using the significance feedback principle to produce multi-layer adjustment. In a typical run it reduced its average error from about 0.2 units to 0.005 in 1000 trials and to 0.00025 in 2000 trials. However, the optimal way in which the feedback principle should operate to determine the changes has not yet been determined.

At present fixed networks are being studied, but their principles can readily be extended to allow the formation of new connections and insertion of new elements. The embodiment of these principles in networks of neuron-like elements should produce highly versatile systems and it would be surprising if these have no relevance to the nervous systems.

References

1. ANDREW, A.M.: Significance feedback in neural nets. Report of Biol. Comptr. Labty., Univ. of Illinois, Urbana (1965).
2. ANDREW, A.M.: "Progress in Cybernetics", ed. J. Rose, p. 359. London: Gordon and Breach: 1970.
3. BARLOW, H.B.: "Mechanisation of Thought Processes", p. 535. London: H.M.S.O.: 1959.
4. BARLOW, H.B.: "Sensory Communication", ed. W.A. Rosenblith, p. 217, New York: M.I.T. Press and Wiley (1961).
5. BEURLE, R.L.: "Principles of Self-Organization", ed. H. Von Foerster and G.W. Zopf, p. 291, Pergamon Press (1962).
6. BRUNER, J.S., GOODNOW, J.J. and AUSTIN, G.A.: A Study of Thinking, New York: Wiley (1956).
7. BLAKEMORE, C. and COOPER, G.F.: Nature $\underline{288}$, p. 477 (1970).
8. BRAINES, S.N., NAPALKOV, A.V. and SVECHINSKI, V.B.: Problems of Neurocybernetics. Moscow: Academy of Med. Sci. (Translation NLL/24/4/BRAI from Natl. Lending Library for Science and Tech., England, or OTS: 60 - 41, 639 from U.S. Bureau of Commerce) (1959).
9. DREYFUS, H.L.: Alchemy and artificial intelligence. RAND Corp. report P 3244.
10. GABOR, D., WILBY, W.P.L. and WOODCOCK, R.: Proc. I.E.E., London, Pt. B., $\underline{108}$, p. 422 (1961).
11. GREGORY, R.L.: Ergonomics, $\underline{13}$, p. 25 (1970).
12. HEBB, D.O.: Organization of Behavior, New York: Wiley (1949).
13. HUBEL, D.H. and WIESEL, T.N.: J. Physiol. $\underline{195}$, p. 215 (1968).
14. HUNT, E.B., MARIN, J. and STONE, P.J.: Experiments in Induction. New York: Academic Press (1966).
15. KOGAN, A.B.: Scientia, $\underline{103}$, Sept. - Oct. issue, p. 1 (1968).
16. MINSKY, M. and PAPERT, S.: Perceptrons. Cambridge, Mass.: M.I.T. Press (1969).
17. ROSENBLATT, F.: Principles of Neurodynamics. Washington: Spartan Books (1962).
18. SAMUEL, A.L.: I.B.M. Jnt. Res. and Dev. $\underline{2}$, p. 320 (1959).
19. SINCLAIR-DE-ZWART, H.: "Studies in Cognitive Development", ed. D. Elkind and J.H. Flavell, p. 315, New York: O.U.P. (1969).
20. STAFFORD, R. (Ed.): "Biophysics and Cybernetic Systems", In: M. MAXFIELD, A. CALLAHAN and L.J. FOGEL, p. 81, Washington: Spartan Books (1965).
21. TOMOVIC, R.: "Sensitivity Analysis of Dynamic Systems". New York: McGraw Hill (1963).
22. UTTLEY, A.M.: Brain Research $\underline{2}$, p. 21 (1966).
23. UTTLEY, A.M.: J. Theor. Biol. $\underline{27}$, p. 31 (1970).
24. WINOGRAD, S. and COWAN, J.: "Reliable Computation in the Presence of Noise", Cambridge, Mass.: M.I.T. Press (1963).

DISCUSSION

ROSEN:
I believe that one aspect of importance for arguments of the type presented involves the fact that our neural apparatus decomposes incoming stimuli into sets of "features", probably in different

ways simultaneously, and provides us with strategies for determining which "feature"is significant in a given context. We then respond to the "feature" and throw away the rest, but this is highly context-dependent; in another context, another "feature" of the same stimulus will be extracted. This is different from just redundancy reduction; indeed, in many cases we will add to the incoming stimulus, as when we hear a harmony under a bare melody, or when we recognize a caricature. It seems to me that such schemes as "significance feedback" are most useful in understanding what happens once a particular context and its associated "features" are extracted, but they leave the basic problems of the association of contexts and "features" untouched.

ANDREW:
I would agree with ROSEN that much of the processing of information by the nervous system depends on extraction of features as he says, and that different features are important in different contexts. There are, however, some kinds of processing which are not readily described in such terms, one being that which enables a player in a ball game (e.g. cricket) to estimate the position and velocity of a ball high in the air and to control his movement so that he positions himself to catch it, with successive corrections as the ball comes nearer. Feature extraction is only one, rather minor, part of this process. - Surely, however, some form of "significance feedback" is precisely what is needed to set up a system which will indicate which features are significant in different contexts. I agree that the forms of feedback I have used to modify the properties and interconnections of individual elements are rather different from the kind needed to deal with features. I have started with what seems to be the simplest case.

YOCKEY:
The remark about the difference between noise and redundancy is very important but I doubt if anyone not familiar with information theory will appreciate it. A cross reference to my discussion of conditional entropy may help. Infrequent patterns carry very little information; i.e., if

$$p_i \text{ is small } H_i = - p_i \log_2 p_i \text{ will be small}.$$

ANDREW:
I thank YOCKEY for his comment; at the time of writing I do not have a copy of his paper available and cannot reply in detail.

PAVLIDIS:
This is an interesting discussion of learning, self organization and artificial intelligence. However, there are a few points which I would like to take exception with. The experience of many workers in the field of pattern recognition has been that "learning" schemes are less valuable than what was originally thought. Better results can be obtained if the input is subjected to elaborate preprocessing followed by a very minimal amount of learning, if any (+). Such techniques are not necessarily linguistic. Their main feature is rather structural analysis accomplished by preselected algorithms (ibid). In terms of psychological analogs these could be characterized as "genetically determined" or if ontogenetic, as the results of much more general experience than the one directly related with the task to be performed. This description seems to correlate well not only with the work by HUBEL and WIESEL (Ref. (13) in the paper), but also with that of BLAKEMORE and COOPER (Ref. (7) in the paper): The organization of the cortical units can possibly be achieved during early infancy but they are ready for the performance of more complex tasks. Thus attempts to train an automaton to perform a complex task through self organization may be overly optimistic. If any ontogenesis is possible it could only come with much wider experience.

+ PAVLIDIS, T.: Technical Report, No. 89, Computer Science Laboratory, Princeton University (1971).

ANDREW:
I think I have no real disagreement with PAVLIDIS. His comment is a timely warning against underestimating the part played by "genetic learning" and also the extent to which living systems carry over something gained from past experience when dealing with new situations. This last point has been discussed by Oliver SELFRIDGE in one of his publications. By way of illustration he compares chess-playing by a computer and by a person. The computer has to be told, or has to learn in the chess environment, that it is good to take the oponent's pieces but bad to lose one's own. To human beings, on the other hand, these ideas come very naturally, being similar to some which became ingrained when they squabbled with playmates over possession of toys at a tender age.

COULTER:
ANDREW's concept of significance feedback based on continuous automatic sensitivity analysis is intriguing. That this would involve adaptive self-organization seems clear, and an ingenious advance on perceptrons. It is not quite so clear how this involves "goal-directed" self-organization, but perhaps this is a matter of definition. - One question left unanswered is: where do the goals come from? Most theories of purposive activity assume the goal as somehow given, and focus on the mechanism by which the goal is attained. This is important, but a theory of the mechanism of goal formation is also important. For a theory of the ontogenesis of purposive activity to be complete, it seems to me a theory of teleogenesis needs also to be included. This is not intended as a criticism, but as a suggestion for further work.

LOCKER:
Ascribing goal-directedness to a self-organizing system presupposes a goal that is either given from outside or has arisen inside the system. In the latter instance the term goal-directedness would express a sort of self-tendency, namely towards assembling itself around an imaginary (or virtual) centre. If the acquisition of language happens without the use of language and, similarly, the acquisition of skill in most cases requires a little verbal content only, then it looks as if 1. the formal structures according to which the process of origination or generation of something occurs are previously given, 2. what originates has little influence on the process of origination itself, and 3. the concretization of something abstractly given depends on a trigger for this process. In particular, learning is often interpreted as realization of something preformed. - Is redundancy reduction akin to a certain kind of system's concentration (and simplification of its diversity) in order to facilitate the self-organization? Your definition of noise is intriguing since it would reveal the prevalence of the system's conditions (and general constitution) over the processing of signals. In as much as the system alters, e.g. during the process of self-organization or adaptation, a signal representing primarily only noise, i.e. being entirely meaningless, yet disturbing, could suddenly assume the role of meaning.

ANDREW:
LOCKER and COULTER both ask how the goals are set in the first place. This is a tricky question and one where it is necessary to be very careful about definitions of terms. COULTER is right in saying it is a major question for further study; it is, of course, also a problem in the context of perceptrons, which operate with the goal of coming to produce response agreeing with those of the teacher. - A glib answer is to say that the overall goal of living systems is the survival of the individual and of the species. The other goals, such as those of obtaining food, keeping warm and dodging enemies, can be regarded as subgoals whose achievement allows achievement of the main goal. A better way of looking at it is to say that we are only able to observe those species which have survived, and these are the ones which have somehow come to be directed towards the goals of obtaining food, etc. - Apart from living systems, crystals have a special type of organization and in the case of certain crystals such as diamonds, this is strongly conducive to survival. The essential difference between crystals and living organisms is that the former behave passively while the latter survive by means which are much more "clever". The difference can be expressed in rigorous terms by invoking the idea of "directive correlation" introduced by

SOMMERHOF (1950). The goals of living systems must be evaluated internally according to the above view and hence it could very well be appropriate to show a "goal evaluation centre" in a functional diagram of a living system. This does not imply that the centre can be located as a physical entity. It is convenient, though perhaps not always justifiable, to talk as though this centre made available a continuous indication of the degree of goal achievement as a numerical value. WIENER (1948) has termed this "affective tone"; SELFRIDGE (1956) has coined the attractive term "hedony". - I have suggested that the goals are internally evaluated, and that much of the self-organization is independent of language. I do not think these suggestions imply that the learning or self-organizing process can only operate to trigger off complex patterns of behavior whose details were previously stored. Behavior is determined both by the "genetic learning" of the species and by learning during the lifetime of an individual. To suggest that learning in the lifetime can only trigger off complex patterns is to expect a great deal of the genetic learning, which also hat to act without the benefit of linguistic reasoning. - I now think that the best way to look at redundancy reduction is as a heuristic device, in the sense in which "heuristic" is used by writers in the Artificial Intelligence field, e.g. NEWELL, SHAW, and SIMON (1959). The goal of redundancy reduction has no simple relationship to the main goals of the system, but does seem to narrow down the search for states of organization which will achieve the main goals. In reply to LOCKER's last point I would reply emphatically "yes". In any learning system worthy of the name, signals which initially have the effect of noise, because the system has not learned to utilise them, can come to be used.

Some General Problems of Memory
J. S. Griffith

Abstract
The existence of human long-term memory poses many general problems. It is shown that a Fourier series provides an interesting analogy for the distributed property of memory and the question is discussed of whether memory is stored in a discrete or continuous parametrization.

1. Introduction

There are many forms of "biological memory". Of these, perhaps the most important are the following three. First there is genetic memory, largely stored in coded form in the base sequence of chromosomal DNA. Secondly we have immunological memory, which is less completely understood but is connected with the ability of an organism to maintain a capability to synthesise protein molecules (antibodies) which are in a physical sense complementary to foreign molecules (antigens) previously present in the organism. Finally there is memory in the everyday sense of the word namely the ability of the central nervous system to be so altered by previous experience that it reacts subsequently in a manner which reflects this alteration. The most dramatic form of such memory appears of course in the ability of man to remember over long periods and then to reproduce long sequences of symbolic material, as lists of numbers or segments of text. In this case, although many speculations have appeared and much experimental work has been done, we must admit that we still appear as far as ever from any certainty as to mechanism. In this paper I shall briefly discuss some of the questions raised by the existence of long term memory in the central nervous system.

2. The Read-in and the Read-out

To understand the storage of information in a computer we must know how it is read into the computer and, subsequently, how it is read out again. A superficially similar question must be asked about biological long term memory. Here I think the problem of thinking of any plausible detailed mechanism is much more difficult for the "read-out" than for the "read-in". For, as was clearly pointed out by BEURLE (1), small changes in almost any cellular property arising as a consequence of nerve cell firing activity could, if maintained, serve as a "memory" of that activity. The most natural possibilities here are, of course, a change in neuronal connectivity as originally discussed by TANZI and CAJAL (see CAJAL (2)) or in neuronal threshold (SHIMBEL, (20), see also ECCLES (4), p. 219-221) although glial cells could also possibly be involved (6-8), (14), (22).

The read-out poses more difficult problems. Partly we have to ask how a given firing activity in the brain, A say, which is the primary representation of an event which is to be learnt and which has led to a certain modification $S \longrightarrow S + \delta S$ of anatomical or functional structure of the central nervous system, could be reconstructed by the brain from this modification δS at some future time. This question can probably only be considered sensibly in relation to some fairly well-defined theory of functional organization of the brain, as for example I have done in my own theory of the latter (10), p. 40 - 41).

However, there is another question which we can consider to some extent independently of any specific model of brain activity. Following the original work of LASHLEY (5), it is generally agreed that although the cortex shows varying degrees of localization of function, individual memories are almost certainly not each stored in separate places (for example one in each neurone). Rather, they are each spread diffusely and largely coextensively over great regions of the cortex. The question then arises of how these intermingled memory traces can be disentangled, the one from another.

A simple and instructive analogy here is furnished by the relation between a function $f(x)$ and its Fourier components (9), p. 431. For example, consider a finite Fourier sine series

$$(1) \qquad f(x) = \sum_{n=1}^{N} a_n \sin nx$$

in which each coefficient a_n is either $+1$ or -1. A knowledge of the a_n, which requires N bits of information, trivially tells us what the continuous function $f(x)$ is. Conversely, if we are given $f(x)$, we can evaluate any of the a_p by the well-known formula

$$(2) \qquad a_p = \frac{2}{\pi} \int_0^{\pi} f(x) \sin px \, dx$$

We have a relation between the discrete components of the vector $\underline{a} = (a_1, a_2, \ldots, a_N)$ and the continuous functions $f(x)$. The effect of setting a particular $a_p = +1$ rather than -1 has an effect on $f(x)$ over the entire range $0 < x < \pi$ and thus the "memory" of its value is spread over the whole of this range. Yet we can easily reconstruct the value using eq. (2).

This relation between vectors and functions is of a central importance in pure mathematics in the theory of Hilbert spaces and in quantum mechanics in the correspondence between the Heisenberg and Schrödinger representations. It shows in principle how it is possible to convert a discrete sequence of symbols into an essentially continuous distributed modification of brain structure and to extract it again from the latter form. Of course, in the brain, the auxiliary functions occurring in eq. (1) need not necessarily be sines or cosines but could really be any linearly independent set of functions $f_n(x)$ although then the extraction procedure would not necessarily be as simple as is shown in eq. (2). Nor need the relation between $f(x)$ and $f_n(x)$ necessarily be linear. Non-linearity greatly complicates the mathematical analysis but is quite likely in reality because of the strongly non-linear character of the input-output relations of neurones.

More recently the physical technique of holography has been suggested as an analogy to this sort of property of memory, and this involves essentially the same sort of mathematical considerations (5), (16), (19), (23). One feature of a hologram is that the original picture can be reconstructed, albeit in a slightly less sharp form, from even a part of the hologram. This has attracted particular interest because it offers an analogy to the fact that quite extensive cortical damage can often have a relatively small effect on memory. Such a characteristic is also present in equation (1) and indeed the values a_n can be extracted even if the values of $f(x)$ are only known in an arbitrarily small interval $x_1 \leq x \leq x_2$. This is easily seen in the following way. Take an interior point x_0 of this interval at which none of the sines in equation (1) vanish.

Differentiate equation (1) 4m times to get

(3) $$f^{(4m)}(x_o) = \sum_{n=1}^{N} n^{4m} a_n \sin nx_o$$

Taking $m = 0, 1, 2, \ldots, N-1$, we have N linear equations to determine the a_n, with determinant

(4) $$D_N = \left| n^{4m} \sin nx_o \right| = \left| n^{4m} \right| \prod_{n=1}^{N} \sin nx_o .$$

But $\left| n^{4m} \right|$ is the alternant determinant on the numbers $1, 2^4, \ldots (N-1)^4$, which is non-zero, so D_N is also non-zero, as all the $\sin nx_o \neq 0$. Hence the a_n can be determined.

Of course, there is no reason why the a_n could not have values other than ± 1. But, because of the threshold properties of neurones, it might be interesting to consider the case in which only two ranges of values of a_n, say $a_n < \theta$ and $a_n \geq \theta$, are regarded as distinguishable. However, it should be clearly recognized that such analogies are not theories of memory, although they do help to make certain postulated features of real memory appear less mysterious to us.

3. Continuous or Discrete Parametrization of Memory

In a digital computer the information is stored in a discrete form and the memory maybe regarded as a series of 0's and 1's. The physical representation of these is typically in the form of a direction of magnetisation which has two possibilities, e.g. round a ferrite ring one way or the other. In an analogue computer, however, the representation is by means of a continuously variable quantity, such as a voltage. In a digital computer the discreteness of the information offers the opportunity of building a stability into the memory and the circuitry, which is not so easy when one has a continuously varying parameter with its consequent tendency to drift.

Which of these possibilities occurs in the brain is not yet known, although a similar opportunity for stability exists there and I think a discrete representation is more probable for this reason. There are at least two rather distinct ways in which this could occur. One is that chemical markers, serving a control function in relation to mRNA and to protein synthesis, might be placed on the neuronal DNA and be the ultimate physical repository of memory. The possibility that long-term memory might exist in this form is rendered more attractive by the fact that nerve cells do not normally divide during an animal's lifetime. This idea was implemented in the DNA-ticketing theory of memory (13) using the ticketing hypothesis of control of mRNA translation of SUSSMAN and SUSSMAN (21).

The other way arises through the property that non-linear equations have of often possessing two or more distinct solutions which are stable in respect of small perturbations in their dependent variables (see for example (18)). The possibility that this might be important in biology has been recognized for a long time (3), (17), and it certainly offers a way in which a cell might, rather like an atom in quantum mechanics, have two or more distinct stationary states not differing at all from each other in their DNA. The process of remembering something would then involve transitions (or switching) between these stationary states. A discussion of this idea has been given elsewhere in some detail (10-12) under the name of a switching theory of memory.

References

1. BEURLE, R.L.: Phil. Trans. Roy. Soc. London A240, 55 - 94 (1966).
2. CAJAL, R.Y.: Histologie du Système Nerveux, 2 volumes, Madrid: Instituto Ramon y Cajal 1952.
3. DENBIGH, K.G., HICKS, M., PAGE, F.M.: Trans. Farad. Soc. 44, 479 - 494 (1948).
4. ECCLES, J.C.: The neurophysiological basis of mind. London: Oxford University Press (1953).
5. GABOR, D.: Nature 217, 584 (1968).
6. GALAMBOS, R.: Proc. Nat. Acad. Sci. 47, 129 - 136 (1961).
7. GLEES, P.: Neuroglia: Morphology and Function. C.C. Thomas (1955).
8. GLEES, P.: In: "Biology of Neuroglia", Windle (Ed.), p. 234 - 242. C.C. Thomas (1958).
9. GRIFFITH, J.S.: In: "Molecular Biophysics", Ed. B. Pullman and M. Weissbluth, pp. 411 - 435. New York: Academic Press (1965).
10. GRIFFITH, J.S.: A View of the Brain. London: Oxford University Press (1967).
11. GRIFFITH, J.S.: In: "Quantitative Biology of Metabolism", Ed. A. Locker, p. 234 - 244. Berlin: Springer-Verlag (1968).
12. GRIFFITH, J.S.: Mathematical Neurobiology. Academic Press (1971).
13. GRIFFITH, J.S. and MAHLER, H.R.: Nature, 223, 580 - 2 (1969).
14. HYDEN, H.: In: "Macromolecular Specificity and Biological Memory", Ed. F.O. Schmitt, p. 55 - 69. M.I.T. Press (1962).
15. LASHLEY, K.S.: Brain Mechanisms and Intelligence. Reprinted (1963), New York: Dover Publications (1929).
16. LONGUET-HIGGINS, H.C.: Proc. Roy. Soc. B, 171, 327 - 334 (1968).
17. LOTKA, A.J.: Elements of Physical Biology. (1925). Baltimore: Williams and Wilkins Co. Reprinted as "Elements of Mathematical Biology" by Dover Books, New York.
18. MINORSKY, N.: In: Dynamics and Non-linear Mechanics" by E. Leimanis and N. Minorsky. New York: John Wiley and Sons (1958).
19. PRIBRAM, K.H.: In: "Macromolecules and Behavior", p. 165 - 186, Ed. J. Gaito. New York: Appleton-Century Crofts (1966).
20. SHIMBEL, A.: Bull. Math. Biophys. 12, 241 - 275 (1950).
21. SUSSMAN, M.: Nature, 225, 1245 (1970).
22. VALLECALLE, E., and SVAETICHIN, G.: In: "Neurophysiologie und Psychophysik des Visuellen Systems", p. 489 - 492. Ed. R. Jung and H. Kornhuber. Berlin: Springer-Verlag (1961).
23. WILLSHAW, D.J., BUNEMAN, O.P., and LONGUET-HIGGINS, H.C.: Nature, 222, 960 - 2 (1969).

DISCUSSION

ROSEN:
I have always been troubled by a problem which arises when we try to identify a "read-out" mechanism in neural nets, when memory is supposed to occur by means of wiring changes or changes of cell properties in the net. Namely, what is to prevent such changes from occurring in the "read-out" portions of the net itself, thereby continually altering our "memory of memories?". To avoid this seems to require postulation of an extraordinary kind of differentiation in the nervous system, stipulating that such cell or wiring changes cannot occur in the "read-out" portion of the net. (The problem does not arise in genetic systems, since an alteration of the "read-out" system there would almost certainly be lethal.) - The remarks regarding the coding of a continuous function by its FOURIER coefficients, and of the applications of holography to neural process described by LONGUET-HIGGINS, remind me very much of some of the learning

models of Steven GROSSBERG. In his case we have a system initially capable of producing all possible "overtones" or harmonics (though in a non-linear setting, rather than the linear FOURIER setting); these overtones can be recombined and reweighted to give any other response. To train the system to give a desired response it is only necessary to couple it to a feedback mechanism which will enhance the desired overtones and suppress the undesired ones.

COULTER:
GRIFFITH has made an important point in showing that holographic theory is not the only way to account for the distributed property of memory. Other ways are also possible. One, for example, would be the recognition of a figure as a member of a category, with further specification by parameters to distinguish it from other figures in that category. A straight line is defined by the fact that it is a straight line, plus two parameters. The economy of coded representation so achieved would permit not only replication and distributed storage but also rapid readout of the original figure from any of the storage points.

FONG:
It is true that the read-out of memory is a more difficult problem than read-in. Yet there is an easy solution: One may assume a memory mechanism that is closely analogous to that of a tape recorder for which the mechanism of read-out is clearly understood and is easily realizable physically. In fact I have developed a ferroelectric recording theory of memory using RNA as the recording tape to record the sequence of nerve impulses representing a sensory perception in a way analogous to the recording of an oscillating magnetic field corresponding to a sound wave by a recording tape with the major difference that ferroelectricity substitutes for ferromagnetism (FONG, Physiol. Chem. Phys. 1, 24 (1969)). In such a theory read-out is not a serious problem. The theory applies to both short term and long term memory with minor modifications. While the theory implies that memory trace is localized on a molecule, the experimental fact of non-localization of long term memory in the brain can easily be explained by the assumption that many copies of the "recording tape" are made at the same time and are deposited at different parts of the cortex so that even extensive cortical damage does not wipe out memory completely. This recording mechanism is in principle an analog recording (of the train of nerve impulses) and thus the complications of digitalizing information and the use of a memory code are eliminated. Any other memory mechanism is likely to involve a read-out process that is difficult to realize physically and evolutionally. It seems to me that more attention should be paid in memory research to explore possibilities along this direction.

LOCKER:
Did you reflect on the restrictions that must be assumed for a mutual transfer of concepts dominating in discrete or continuous systems, respectively, and for which ROSEN has introduced a "Correspondence Principle "?

GRIFFITH:
(Final remark not received)

Role of Glycoproteins in Neural Ontogenesis, Membrane Phenomena, and Memory

E. G. Brunngraber

Abstract +

The structure of membrane-bound glycoproteins is briefly discussed. Glycoproteins are associated with the cell plasma membrane. Their location on the cell surface has led to suggestions that the glycoproteins may have a directing influence on the establishment of the correct cell to cell connections in the developing nervous system. Glycoproteins may be associated with the serotonin receptor, act as a receptor for other transmitters, and may play a role in ionic exchanges at the cell surface. Short-term memory is believed due to rapid conformational changes in the nerve cell membrane, especially in the synaptic region. Long-term memory is believed to be a consequence of the permanent alteration of the membrane due to deposition of glycoprotein material in the cell surface as a consequence of neural stimulation. The glycoproteins thus deposited alter the properties of the membranes. Membrane structure and properties may also be changed by the alteration of the structure of the carbohydrate portion of the glycoproteins as a consequence of neural activity. Alterations in the spatial arrangement of glycoproteins on the neuronal surface may also influence subsequent membrane phenomena. Reference is made to the possibility that turnover of nerve terminals may provide a mechanism of information storage. Difficulties in demonstrating chemical mechanisms of information storage are considered.

Reference

BRUNNGRABER, E.G.: J. Pediatrics 77, 166 (1970).

DISCUSSION

GABEL:
It has been found that mitochondria contain a protein-synthesizing system complete with RNA and DNA. May not the plasma membrane also contain such a system? If the membrane can synthesize protein, then may not the difficulties inherent in the delay due to transport by avoided?

BRUNNGRABER:
As a matter on fact, there have been reports that purified membrane preparations from HeLa cells and liver cells contain small amounts of RNA. DNA has not been demonstrated in plasma membranes. Unfortunately, it is difficult to know at this time whether the small amount of RNA is actually present in the membrane or in an impurity. Let us assume that membranes do contain RNA. Experimental data suggests that plasma membrane RNA represents only a very small percentage of the total RNA of the cell, and the contribution of this type of RNA to the total protein synthesis must be negligible. Furthermore, to avoid the delay due to transport, the plasma RNA must be capable of operating to synthesize altered proteins without the operation of messenger RNA and DNA from the nucleus. This would be a unique protein-synthesizing system. In

+ Full paper to be published elsewhere.

our calculations, we estimated that only 5 molecules of a protein containing 100 amino acids can be synthesized per synaptic knob during the time of an impulse. The assumptions upon which this calculation is based are such that this figure probably exceeds, by far, the actual synthetic capacity of the cell. Some idea of the rate at which polypeptide chains increase in size was reported by T. HUNT, et. al., (J. Mol. Biol. 43, 123, 1969) who calculated that the growth rate of hemoglobin chains proceeds at about 10 amino acids per second. It therefore takes 0.1 second to add one amino acid, while neural events are measured in terms of 0.001 to 0.002 second.

FONG:
It is interesting to learn that short term memory is not likely to be of a chemical origin. The theory of memory I proposed (Physiol. Chem. Phys. 1, 24 (1969)) is based on a physical change of the conformation of a macromolecule. With a slight modification the mechanism may be applied to long term memory as well; therefore long term memory need not be basically different from short term memory. The theory I proposed can be tested by experiment; see recent work of STANFORD, et. at. (Physiol. Chem. Phys. 2, 499 (1970)).

BRUNNGRABER:
This is, of course, possible. However, one must take into account the fact that most substances in brain turn over metabolically. Assume that the neuron synthesized macromolecule A from its precursors (reaction 1). The conformational change responsible for long term memory involves a change from A to B (reaction 2). We have the sequence:

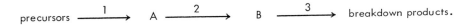

precursors $\xrightarrow{1}$ A $\xrightarrow{2}$ B $\xrightarrow{3}$ breakdown products.

If B is continuously breaking down (reaction 3), it is difficult to see how the conformational change would provide an adequate mechanism for long term information storage. However, reaction 3 could be exceedingly slow. Certain proteins in brain, in fact, do have a very slow turnover. We would expect therefore that B would accumulate with age. Accumulation of specific substances with age has not yet been found, but this does not exclude the possibility of finding such a compound as more sophisticated methods of isolation and determination are developed. Neurochemists, interested in information storage, are generally more interested in looking for substances with a rapid turnover. I have often wondered whether the substances of very slow turnover may not be of more significance. - The chemical theory, which I favor, has the advantage over the foregoing model, since it allows one to avoid the difficulty due to metabolic turnover.

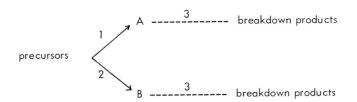

Before learning, the cell produces A. During the process of memory consolidation, the cell ceases to produce A, but produces B instead. In this model, tunrover rates of A and B would be irrelevant. It is well established that dramatic chemical changes occur in stimulated neurons and ultimately it may be possible to demonstrate a chemical difference between cells that had learned and those which had not. - In my article, I referred to the possibility of chemical and morphological changes in the Golgi apparatus (which is the subcellular organelle responsible for the biosynthesis of glycoproteins) and that such changes may produce altered glycoprotein molecules. A recent report (W.O. WHETSELL, Jr., and P.R. BUNGE, J. Cell. Biol. 42, 490, 1969) provided evidence that the Golgi apparatus of sensory neurones swell when these are stimulated by ouabain.

On Controlled and Totally Neural-Replies Generated Concepts for Biology and Functional Brain Theory

D. L. Szekely

Abstract +

The control of concepts and conceptual coherence in biology and functional brain theory has to be extended by dimensional analysis. Controlled concepts have to be applied within closed and verifiable methodological circuits. Dimensional analysis has to be extended to various of its identity cases. The generalizing extension requires a new logic having a heterogeneously interpreted categorization of physical basic domains as its foundation. Central to this logic is the concept of heterogeneous coordinative relation. The scheme of this relation generates a physical neuron and a physically generalized Turing machine. The theory of the total reduction to a single physical ultimate demonstrates how physically important scientific key concepts are to be generated: either by a physically back-connected neuron or by the physically generalized Turing machine. As the human brain applies physical neurons by means of the above mentioned method concept generation and transformation can be simulated: but also the concept generating possibilities of the human brain can be optimized and the process of programming brain-like systems can be explained.

The Menace of Antinomy if a "Brain" is thinking about "Brain"

A brain theory cannot be erected without a few thoughts of somewhat heretic nature in that they are not made in a style familiar to the professional logician. A logician knows that at an abstract level a functional value of a certain function cannot be inserted as an argument in the same function. An antinomy would inevitable arise and the general requirement to discard just that symbolic structure or system had to be followed. The theory of content-transmitting languages did not reach yet that very level of formalization which would permit analogous thoughts for such languages. Our content-transmitting languages are neither fully controlled nor fully abstract structures. They have no axioms; only crude approximations of what axioms should be. They can be regarded as a set of interrelations of functionally perceived approximations of axioms and are strongly influenced by their semantically coordinated interpretations., Logicians tried to dissect the constellation in which a language is speaking about another language: in the accepted terminology it is a "meta-language" which makes assertions about an "object-language". Unfortunately, without stringent restrictions and affixed compensatory rules, this relation leads - after several repetitions - to an infinite divergence which is as bad as any antinomy. This explains why we need a new basic approach.

The actions of the human brain at a conception-conscious level are dependent on some language or code. If a language cannot speak, in general, about another language without involving the danger of antinomy, if, in addition, a language cannot speak about itself in a logically safe manner - then we must raise the question: can a brain think about itself in a logically correct way? Which restrictions and compensating operators are necessary to exclude the danger of antinomy? How can we exclude such limiting effects on the range of our thinking capacities?

+ The full paper, containing notes on brain theory, on the individuality of the brain and the interindividuality of language and schemes for extended dimensional analytic systems etc., will appear elsewhere.

We can go another step further: if this consideration refers to the specific biological instrument called "brain" — what are the projections thereof on our efforts to introduce conceptually controlled numerical thinking and coherent systems of concepts, to biology in general and brain research in particular?

We may like or dislike the described logical situation — but it exists in some form and is fateful for our efforts to update biological efforts. With reference to the remarks made by ECCLES: perhaps, after all, not the brains of the brain reserchers are poor, but rather the brain conditioning languages are introducing an inefficient linguistic logic. If this is so, indeed, we have to replace present-day linguistic logic by a rationally based physical logic and a corresponding control theory.

References

CARNAP, R.: Introduction to Semantics, Harvard Univ. Pr. 1947.
SZEKELY, D.L.: Proc. 6th Int. Cybern. Congr., Namur, 1970.

DISCUSSION

GRIFFITH:

With reference to the possibility of antinomy occurring when "brain" thinks about "brain", I think it is very important to distinguish between two situations, in one of which an antinomy, or something similar, may occur but probably may not in the other. These situations are the analogues of the two sentences "This very sentence is untrue" and "That other sentence is untrue", of which the first is an antinomy but the second is not by itself. — There seems to me no reason to expect any special limitation of a "self-reference" kind to arise in our attempts to produce a normal scientific type of theory of the brain (as for example in ref. 1). Does one believe there will be such a limitation to our understanding of the biochemistry of E. coli just because our biochemistry has much in common? Or to our understanding of the neural basis of behaviour in a fly because its neurones may well be quite similar generally to our own? Why, then, should a self-reference problem arise when my brain seeks to understand the neural basis of "that other man's" behaviour, just because his brain is made of nerve cells like unto my own? — The antinomy-like situation occurs when my thoughts turn inwards and ask questions like "Why is that I feel that I am an integrated entity existing somehow independently of matter, when really my brain is just a collection of nerve and glial cells?" In this sort of context I have the feeling, in common I think with most people, that there are infinitely deep and puzzling questions about the "mind in the machine" which I cannot even formulate in a very satisfactory way, let alone even hope ever completely to answer. I think it is here that the antinomy is obtruding. — This matter may also be put more symbolically by enumerating all brains as B_i, $i = 1, 2, \ldots$ Then the sentence "the brain B_i thinks about the brain B_k" can only give something like an antinomy when we set $k = i$.

SZEKELY:

At a conceptual level the brain can think only by using the formal apparatus of that "brain-programming language" by means of which it has been preprogrammed. A good brain-programming language is adapted to the structure of the human memory system and has fundamentally different structure from all known linguistic devices. — The question how brain B_i thinks about another brain B_k and its components must be scrutinized on the basis of the relation of two programming languages: L_i used for B_i and L_k used for B_k. In general $L_i \neq L_k$. But even if $L_i = L_k$, we should not assume that B_l acquired and memorized during its preprogramming the same "range of programming" as B_k. But intercommunication requires a common section of the range of preprogramming for all members of a linguistic community. — The assertion "B_i thinks about B_k" must be reformulated as "The programmed range of B_i by L_i has the task of a metalanguage in relation to the programmed range of B_k by L_k. The two ranges (and their structural content) are interrelated to a meta-relation. To avoid the antinomy of infinite regressive divergence, for the same designatum B_i has simpler structures than B_k. Quite often B_i applies abbreviations where B_k has full structures. The meta-language L_i "thinks" about the objects designated by L_k by means of such simpler structures and abbreviations. B_i prefers generality and disregards structural details of lesser logical importance. Thus, B_i and B_k apply different local ranges of the same programming language by giving preference to different type-levels of descriptive-designative patterns. In simpler terms: B_i "thinks" by means of more general patterns and their abbreviations about the more specified structural patterns of B_k. – A neuron cannot think, it can only react. A not-yet programmed brain cannot think at a conceptual level. – A long way of constructive steps leads from the elementary neural-reply to the interpreted calculus used as a brain-programming language. – Thinking and meaning are dependent on the total programmed brain-system which is a compound structure described by concepts of a very high level of types. Problems of self-reference arise only at very high "system"-levels. Self-reference is not a neural problem but that of a degenerated intersystem-relation. – It seems to me that in the case of self-reference we face a temporary interchange of the meta- and object-tasks. This is a methodological step which, if not carefully compensated, can lead to new variants of antinomies. The same exchange of the meta- and object-tasks within the schema of the same metarelation seems to be applied whenever B_i thinks about such details of B_k which are too specific for the generalizing abbreviations of B_i. The results must be retranslated later into the original form of the metarelation.

BREMERMANN:
If anything this paper proves the futility of a purely linguistic approach. I doubt that the practicing biologist is familiar with the antinomies of logic and with the divergence of sequences of metalanguages. This paper does little to explain these difficulties nor how they are related with the problem of a brain trying to understand itself.

SZEKELY:
The dominant idea of my paper is the total rejection of the presentday futile linguistic methods for the conditioning of the human brain. What we need is a re-conditioning of the brain by new systems of coherent concepts controlled by a) a generalized dimensional analysis, and b) concepts which are subject to a second control by closure oriented methodological circuits and their metalogical scheme. Such circuits include the l o c a l application of abstract systems of concepts, their interpretation, – resulting in compound physio-logical systems – and rules developed from heterogeneous relations coordinating to basic physical units. Not the abstract neuron of J. von NEUMANN, but only the task-specified, domain-coordinated, therefore p h y s i c a l n e u r o n , can generate the more efficient concepts. Such "well-generated" concepts are 1) subject to the detailed c o n t r o l by the exponential functions of a generalized dimensional analysis, and 2) can be regarded (in their memory-address section) as derived dimensional units generated by the fired – not-fired patterns of this physical neuron.

LOCKER:
It is true that concepts arise within the realm of logic which can be considered as physical logic, if it exceeds the purely formal realm and assumes contentual reference. But logic refers, in its nature, to a system, i.e. subject-like entity, which can, of course, be transformed or transposed, if appropriately controlled. I am not sure whether in your idea of concept construction this systems theoretical viewpoint is sufficiently considered. The appearance of antinomies must not be the reason to discard the underlying structure as invalid, but rather – at which situation GOEDEL's theorem could be a hint – demonstrates that a more universal concept had to be introduced with respect to which the (original) antinomy vanishes. Respecting the so-called objectively given reality ultimate assertions assume by necessity the form of antinomies; a fact incorporated also in many-valued logic (and ontology) after Gotthard GÜNTHER. A seemingly true antinomy is reached, when one speaks of concept construction by the neuron (i.e. the system) and

the neuron is simultaneously the result of the constructing concept, amounting to the opposition of construction and recognition, or passive mapping and active seizuring of (conceived) objects. Therefore, your idea of an interindividual programming of the individual brain by language can be interpreted as a self concretization of the formal (principally infinite) structure in as much the brain shares these infinite structures (and relations) by recognizing and (re)- constructing them. An interesting point is made by your remark that the physical neuron is intended for reacting with (at least?) 2 data taken from different (linguistic) levels. Here, you arrive at a position where your neuron obviously displays a subject-analogue nature in unifying different levels of the pre-dated structure. Generation (construction) and (re)-construction also seem to be two opposing aspects of the one and the same thing, which - I recall CHOMSKY's hypothesis of innate ideas - is in the ultimate respect consciousness.

SZEKELY:
Physical branches of science apply total methodological circuits made up of fundamentally different sections. One of these is the application of abstract logic both, at elementary and at evaluating levels. Input data for application are prepared and supplied by means of heterogeneous coordinative relations and coordinative rules. In the case of input they are used in a direct form, in the case of output and its evaluating "verification" in a conversed direction. The coordinative rule introduces together with the elementary data the c o n t e x t for physical data. The logical mechanism for the processing of such context-related data must simulate to some extent the interplay of physical rules or the "context". To this procedure we refer by the expression "physical interpretation". This is essentially a meta-level action. "Methodology" is a metalogical s c h e m e . But the abstract system interpreted by the data of a physical context is not yet a "physical logic". It is no more than the coordination-generated interplay of an abstract and a physical system. A logic deserves the meta-predicate "physical" only if it is applied within a total methodological circuit. This circuit includes the definition of the basic physical domains, their mutual heterogeneity, the rules and aids to measurement and the metalogical c o n t r o l of the total process by means of dimensional analysis. The participation of the interpreted logical system in a complete methodological chain justifies the application of the term "physical logic". Within this coordination-generated background even the "physical ultimate" or the simplest building-block of physical concepts must be at least a "hetero-pair": a pair of data taken from two heterogeneous domains (one of which can be the abstract basic, set-theoretical domain).

Remarks on Mathematical Brain Models

R. E. Kalman

1. Introduction

In this note, the question to be posed is: <u>How can mathematics contribute to brain research</u>? "Mathematics" will mean here exactly the same thing as it does to the contemporary professional mathematician, namely theorems, methods of proof, natural constructions. We will not examine in this paper questions involving data analysis, statistics, numerical formulas, and other pedestrian (though often very useful) things which to the layman are also "mathematics".

We pose the question because of our optimistic prejudice that mathematics <u>will be</u> significant in brain research, eventually. The only rational argument supporting such a position is via analogies. Each developing and "hardening" science (chemistry, physics, economics, psychology, ...) gradually reaches a stage beyond which progress is impossible without organized knowledge, because the implications of existing knowledge are too complex to digest without the help of abstraction, that is, without mathematics. In physics, this situation occurred well before the end of the 19th century, and perhaps even by the middle of the century (Lagrange, Laplace, Hamilton) We do not wish to start dreaming in print as to how and in what form this evolutionary stage will finally emerge in biology. We shall merely try to delimit what might be expected to happen, from the point of view of the nature of mathematics. Briefly, our position is that mathematics will become important for brain research only after we shall be able to describe in detail the highly complex workings of the brain and will need mathematics to make more sense out of what we already think we know.

In any case, if mathematics enters brain research at all, in my view, it will happen first through the formal study of brain models.

2. Nature of Mathematical Models

It may be argued that the development of mathematics has been determined far more by historical accidents than by the needs of (applied) physical or biological science. "New mathematical knowledge takes a very long time to develop. This being agreed on, we might as well accept mathematics as it is at the moment and see what we can do with it, with regard to building brain models. We have essentially two choices, between

a) <u>topological</u> models and

b) <u>algebraical</u> models.

The development of topological (or geometrical) models is based on the desire to capture global features of this situation by slurring over all local details. Thus <u>topological models are locally vague but globally rigid</u>. As a result, such models are, intrinsically, incapable of elucidating or describing the internal mechanisms of nerve-cell behavior or even of saying something of direct biological relevance about the operation of small clusters of cells. But it is possible (we should hope, probable) that certain great overall aspects of brain behavior can only be understood from the topological point of view. (Example: Some mathematical economists would claim that the "law" of supply and demand is no more- and no less - than a particular manifestation of the BROUWER fixed point theorem, which is a typically global, all-embracing statement. Nonexample: In physics, in spite of a highly sophisticated mathematical apparatus, there are surprisingly few cases where global mathematical statements are useful.)

The outstanding exponent of the topological point of view in recent years has been the differential geometer and topologist René THOM. He views the world of changing events as smooth transitions (as in classical mechanics) linked by discontinuouities or "catastrophies". This may well be among the major wisdoms produced by the scientific thought of this century. See THOM (7), (8). On the other hand, to construct anything like a testable model of the brain in the spirit of THOM seems to me to be a hopelessly ambitious task at this time. THOM's theory has had little influence on biological research so far (of course, this is not THOM's principal aim), and I can see such influence developing in the next 5-10 years on fuzzy, philosophical level only . This is a bit discouraging, and especially so because the power of mathematical thought (that is, techniques of proof) can have no impact on the basic problems of the brain until a close and natural connection is established between THOM's new world-view and factually observable phenomena in the brain. Here is surely an area where progress must come, first of all, from outside of mathematics. (I would like to insist in passing on some personal prejudices concerning the dream of topologists: That by neglecting all details-which are messy, or hard to ascertain, or simply unknown - one can still make progress toward discovering scientific truths. This view, which came into prominence around 1910 - 1930 when topology was a young field, is simply not borne out by the facts of scientific research since then, in my opinion.)

In many ways algebraical considerations are exactly opposite of the topological ones. (Even more: one might perhaps claim that each subject is in opposition to the other.) It is a fact that <u>algebraical models are locally rigid but globally vague.</u> Local rigidity arises from the way algebra is built up: We begin algebra by defining certain operations (addition, multiplication, scalar product, vector product, etc.), which cannot be modified later unless we want to begin again from scratch. Algebra works by "building" (inducing) new operations or structures from old ones in a "natural" way, that is, a minimum of arbitrary new assumptions. The results of repeatedly erecting new structures on top of earlier ones is, at least to the nonexpert, quite unpredictable from the starting axioms. (Example: The "modern" algebra in the 1920's flourished by exploiting this process of repeated natural constructions.)

It would be very difficult to assess the influence of algebraical thought on scientific evolution. In certain aspects, this influence has been decisive; in others, algebra has never been tried. (Example: It is, basically, an algebraical fact that the local behavior of regular (or "smooth", "nice", "classical") physical systems is <u>linear</u>. Linearity is a rigid concept; there are many unrealized hopes about "slightly" changing linearity to better conform to the real world. This idea does not work; by going away from linearity, one loses much and gains little. Important success, such as in quantum mechanics, are due to a large extent to the algebraical exploitation of linearity, via group representations. A big and active part of modern system theory is concerned with deep properties of linear systems. Nonexample: Other rigid algebraical structures, for instance, homology theory, the machinery of algebraical topology, have had only minor impact on science so far.)

Presumably algebraical models can be relevant to brain research only if we can establish natural and true connections between some elementary level of neuron behavior and a rigid algebraical setup. In other words, we cannot even hope to use the power of algebra until the details of "signals" emitted by the individual neurons, "coding" of neuronal messages, "logic" and "redundancy" of synaptic connections etc. are understood. The common awareness of and agreement about the existence of such problems has inspired quite a bit of speculative research, but it would be foolish to pretend that the results we now have require algebra for their better assimilation. Still, it might be useful to bear in mind that, beyond slogans such as "signals", "information", "coding", "redundancy", etc. (which were borrowed from the childhood of computer engineering) there exist clear and perhaps more sophisticated concepts, still to be discovered. As a matter of fact, some of these have been discovered in the evolution of mathematical system theory. Some conceptual ideas relevant to this discussion have been presented briefly in KALMAN (6).

What, then can we conclude about algebraical models? Clearly their applications cannot get off the ground without rather specific inputs about the information handling mechanisms built into the neurons (perhaps even requiring going down all the way to the level of molecular biology). Nevertheless, the purely mathematical consequences of an "algebraical" system theory, which have been well explored in the linear case and begin to emerge from the fog in other cases as well, may be of as much biological relevances as the grand world-pictures available through topology. Being aware of this might shorten the unknown amount of time remaining before algebraical models can be advertised as a working tool in brain research.

For instance, as sheer speculation we could ask: Just as the physical world is locally linear, is there (or are there) corresponding local algebraical structures in the biological world?

3. Some Existing Models

If we are willing to allow <u>variable coefficients</u> in our classical models (lumped mechanical systems, electrical networks, computers with variable programs), then the resulting behavior can be extremely diverse. This is surely a very well known fact.

Now, the modeling of the brain might begin by attempting to simulate intelligence. Many early papers (à la McCULLOUGH-PITTS) begin with the equation

(+)   intelligence = very complex behavior;

models are then proposed and accepted or rejected on the basis of whether or not they are capable of "very complex" behavior. But, unfortunately, the fact that a given system with variable coefficients can pass such a critique is utterly trivial. (Example: In the fields of artificial intelligence and adaptive systems, a vast amount of (wasted) work in the 1960's has not brought us closer to knowing how to assess more insightfully the problem of "very complex" behavior. Example: Almost any conceivable system could be built with such components as a modern large-memory digital computer; but does this possibility get us closer to understanding problems which are "very complex"?)

It is sobering and downright discouraging to examine existing brain models and find that they have not progressed substantially beyond the stage represented by the idea of a system with variable coefficients. In models of CAIANIELLO's type (1), (2) the variable coefficients are represented by threshold, as they are, presumably, also in the neuron. Learning takes place by change in the threshold levels; the instantaneous state of intelligence of such a machine is represented by the vector of all its threshold levels. This is an awkward mathematical structure and it is difficult to investigate its properties in depth, perhaps because the setup is mathematically basically "unnatural".

The family of learning models due to GROSSBERG (3), (4) postulates a more elaborate (and more classical) mechanism for learning which incorporates linear weighting parameters and delays in addition to thresholds. The linear parameters are governed by an auxiliary set of equations into which the dynamic explanations of the learning phenomena are to be encoded. (The evolution from the CAIANIELLO to the GROSSBERG models has similarities with the evolution of the theory of adaptive systems.) The GROSSBERG models are rather plastic; they can be adjusted easily to take account of new knowledge about the neuron. Precisely for this reason, however, they are less rigid, less algebraical, and perhaps contain less biological truth than the CAIANIELLO model.

Common to both models is the following strategy of critical evaluation. One asks questions concerning the ability of the model to have memory, exhibit learning, forget, adapt, etc.; that is, by using rough idealizations of certain simple types of behavior exhibited by a living brain. The answer to such an inquiry would consist in the result that models passing such tests exist, together with some indications as to how a particular model exhibiting a fixed type of behavior may be actually constructed. That is, the contemporary theory works by trying to show that a given class of models has the potentiality of a kind of behavior which belongs to the list of intuitively brain-like attributes. What is generally not done, at least not in the theory of the CAIANIELLO and GROSSBERG models, is showing that a single model can simultaneously exhibit several different types of behavior, or that the submodels which account for each distinct behavior type can interact with one another in a nonarbitrary way.

We purposely avoided mentioning here those models (e.g., for the reticular system) which are intended as computer representations of anatomical data. In such cases progress depends on the ad-hoc judgments on the part of the modeler concerning the functions of anatomical units. Such models have only experimental or simulation interest. (Ask the computer to do certain things and compare the result with what happens in the brain.) They cannot be used as a starting point of mathematical study and, in fact, they have no mathematical structure.

4. Critique of the Models

Experience with investigation of the type discussed in the previous section has not been spectacularly fruitful, either in biology or in system theory. In the latter field, which goes back at least to the early 1930's, the conclusion has become entrenched that the study of particular systems is not relevant except in very simple cases (small systems). There is agreement that we need general principles of how to construct classes of models with the property that the models can be made quite specific when further data becomes available.

To illustrate this point, which might seem a bit abstract, let us consider the question of controllability. When formulated in mathematical terms, controllability is a technical (necessary) condition without which control is obviously impossible. Roughly speaking, the assumption of controllability alone is sufficient to lead to the solution of many of the standard problems encountered in control theory. The property of controllability is basic, and it can be expressed and theoretically studied without reference to a numerical specification of the model. When the numerical information becomes available, we can use it to decide if a specific given system is controllable or not. What is especially important here is that the whole problem of control can be brought within the grasp of a global mathematical setup. Incidentally, this particular setup happens to be much more algebraic than topological.

On the other hand, neither the CAIANIELLO nor the GROSSBERG have given rise so far to a general understanding of the possibility of learning; there is no such thing known at present as "teachability" or "learnability". In other words, one can demonstrate the possibility of learning in such models by specific examples, but no global features of these models are known at present which explain the inherent reasons for the learning capability (see, for example GROSSBERG (3)). This is quite different from the situation in control theory; there we know, before the start of the solution of a control problem, that control is possible when the property of controllability holds;

therefore the precise system configuration that results after the details of the control problem are worked out is only of secondary interest. Until we know what are the intrinsic properties of a model which enable it to learn, a serious mathematical study of the model will always be handicapped. It is not enough to know that learning can take place in some examples, constructed with hindsight. Looking at special cases is interesting, of course, but it is no substitute for a general methodology.

To put the issues even more bluntly, let us recall the daring of MENDELEEV in proposing the period table (<u>before</u> the discovery of helium) and exposing himself to "destruction" if helium, when discovered, turns out to be other than predicted by the period table. The models we have been discussing here cannot be exposed to destruction because they make only statements about what <u>can</u> be done (and even these statements are semiquantitative at best). The principal <u>test of a model</u> should be in sharply quantitative <u>statements about what cannot be done</u>. To make this point obvious, let us assume that we could associate with the GROSSBERG model quantity (that is, a set of well-defined mathematical objects into which we can substitute numbers in any given concrete case) which expresses the possibility or impossibility of learning (a "learnability" quantity). Then we could compute the value(s) of such a learnability quantity for any concrete model and check the value so obtained against a learning experiment. We cannot do this now, but if we knew of such a quantity we would have a far better test for our models than any of the "Gedankenexperiments" so far described by GROSSBERG. So a search for such "quantities" could be taken as a reasonable short-term goal for mathematical brain research.

5. Speculations

Returning to the competition between topological and algebraical models, perhaps it will be useful to recall one of the situations, now classical, in quantum mechanics. Consider the 3-dimensional rotation group and its <u>linear</u> representations. It is a mathematical fact that each such representation can be made unitary (by suitable choice of an inner product), and it is a further mathematical fact that the irreducible unitary representations depend on a system of half-integers (fairly complicated). It is then a physical experimental fact that these representations happen to agree with the spectroscopic data from which the early quantum theory developed.

To discuss the possible biological usefulness of a topological vs. an algebraical "philosophy" of mathematics, the preceding facts can be rephrased as follows. We make an algebraical assumption (linearity) but we apply it (to the "nonlinear" rotation group in 3-space) indirectly (via linear representations). To get the consequence of this assumption requires a lengthy, deductive mathematical development (mainly algebra). Eventually we get certain specific numbers (describing the possible irreducible representations). These numbers are a global consequence of the (rigid) local, algebraical assumptions but are not predictable from them in an intuitive way. Agreement with experiment is obtained first on the global level, without specifying in detail how and why any given representation is to be matched to experimental data. Later it is still possible to redescend to the local level, put numbers into a particular representation (without new theoretical assumptions and without much new machinery), and make direct and quantitative comparisons with the experimental situations of greatest interest. The impossibility of regaining the local level for the purpose of detailed verification is an intrinsic feature of topological models (see especially THOM (6), § 5, p. 333); this difficulty alone should suffice to prejudice us in favor of the algebraical point of view.

An amazing (but recurring) fact in physics is that the rigid linearity hypothesis, carefully applied in an indirect way, can say significantly new things. This is true even in the discrete world of elementary particles, as articulated in the snobbish phrase that "an elementary particle is simply an irreducible (<u>linear</u>) representation of "something"." Recall the excitement concerning the representations of SU(3) in the mid 1960's.

Can we really exclude the possibility that also in the biological world there exist some (one or several) rigid local algebraical assumptions?

References

1. CAIANIELLO, E.R., DE LUCA, A., RICCIARDI, L.M.: Neural networks, reverberations, constants of motion, general behavior, pp. 92 - 99 in Neural Networks, Proc. Ravello Summer School, 1967 (ed. by E. R. Caianiello), Springer (1968).
2. CAIANIELLO, E.R.: Outline of a theory of thought-processes and thinking machines, J. Theor. Biology $\underline{2}$, 204 - 235 (1961).
3. GROSSBERG, S.: Embedding fields: a theory of learning with physiological implications, J. Math. Psych. $\underline{6}$, 209 - 239 (1969).
4. GROSSBERG, S.: Neural pattern discrimination, J. Theor. Biol. $\underline{27}$, 291 - 337 (1970).
5. KALMAN, R.E.: On the mathematics of model building, pp. 170 - 177 in Neural Networks, Proc. Ravello Summer School, 1967 (ed. by E. R. Caianiello), Springer (1968).
6. THOM, R.: Topological models in biology, Topology $\underline{8}$, 313 - 335 (1969).
7. THOM, R.: Morphogénèse et stabilité structurelle (Benjamin, New York, 1971).

DISCUSSION

PATTEE:
It is very helpful to read a mathematician's informal as well as non-formal views of the role of mathematics in science. KALMAN's characterization of topological models as locally vague but globally rigid, contrasted with algebraic models which are locally rigid but globally vague, suggested to me that the origin of controls might be expressed as the appearance of topological constraints on algebraic models. I don't believe that the failure of the "topologist's dream" of finding truth by losing all detail is only KALMAN's "personal prejudice". To ignore all detail is to ignore experience. On the other hand, as I have argued in this volume, the selective loss of detail is what establishes classifications, and hence simplifications, of the local, "algebraic" complexity. - It is important to realize that even though algebraic models are locally rigid, as in physical equations of motion, they are seldom constrained enough to execute coordinated functions. As I pointed out, since equations of motion are totally rigid locally, this means that the concept of constraint must be represented by an alternative global, topological model. - The difficulty I see with topology is that even though it is formally deeply involved with algebraic notations, it is conceptually too isolated from algebraic models. There is no natural way to express a topological property as a constraint on the algebraic detail without drastic results (i.e., contradiction, undecidability or modification of the axioms). For example, in THOM's differential topology the effects of "catastrophes" are either rigidly deterministic, or, where there is symmetry-breaking, totally undecidable. Physical controls on the other hand are dissipative constraints, hence they are essentially statistical in nature. That is why statistical mechanics follows dynamics in the hierarchicy of physical descriptions. But unfortunately algebra, statistics, and topology are usually treated as entirely distinct languages which we can speak only one at a time. It is my view that one-language models are inherently too narrow for modelling those biological processes that are essentially hierarchical in nature - and this includes a lot of biology, from the origin of genetic controls to the controls of the brain.

SZEKELY:
The basic fact that the system "brain" is a physical system and that an efficient brain-simulator must apply a physico-methodological logic does not exclude the possibility of mathematical brain-models. Efficient mathematical modelling must be preceeded by models incorporating physical metalogic and methodological control. "Methodology" in general is not fully abstract as

it includes non-abstract stages: it is in general "physical". The new concept must be created against the background of a physically interpreted heterogeneous logic. One of the possible 3 interpretations of an elementary metalogical scheme for (asymmetric and heterogeneous) coordinative relations is the "physical neuron". It is the ultimate generator element of the functional theory of concept-construction by the system "brain". The second and instrumental interpretation is the basis of the physically generalized TURING- machine: i.e. the heteropair-reading TURING-SZEKELY machine. Two or more interpretations of the same abstract system, i.e. isomorphisms are the key to the simulation of the concept-generating and transforming actions of the system "brain". Fully abstract, especially algebraic models for the human brain will be possible only after the successful development of a physico-methodological formalization (and simulation) of the concept-construction actions of the human brain. - A well designed brain-programming language, based on a full generation by the mentioned physical neuron, is able to fill this gap. After all, an axiomatically conceived and scientifically interpreted, well designed code is just a third, a semantic interpretation of the same isomorphism. The abstract kernel of such a code can lead the algebraist in his construction of a totally abstract model for the system "brain". But I fully agree with KALMAN: the algebraic model will be efficient only with the interpretative "help from outside", the logic and techniques of a physical environment, - formalized by a brain-programming code approaching in its structure the isomorphism mentioned.

LOCKER:
I think that the difficulties you describe cannot be overcome in principle since a universal mathematical theory that encompasses local as well as global views of equal stringency cannot exist for several reasons. But it should be pondered on the correspondences between the two kinds of view with respect to concrete problems to be solved, as you correctly propose. In other words: although it seems plausible that the progress in details is favored by local considerations only, theories must always remain to be global (in the broadest sense of the term) by means of necessarily neglecting details. However, so far as the problem turns out to be determined subjectively, i.e. by the manner of focusing or viewing it, it probably remains to be an ever lasting alternative that inevitably must be held in an aporetic pending. Nonetheless, correspondences like those between "teachability" on the one level and "controllability" on the other, which you postulate, should be worked out carefully.

KALMAN:
(Final remark not received)

How to Conceive of Biogenesis (A Reflection Instead of a Summary)

A. Locker

It seems evident that every problem and also that one dealt with in this volume can be considered in a twofold way; namely quasi-objectively, tending to neglect the observer, and subjectively, emphasizing the role of the observer in the description. For both modes of view this volume provides excellent examples. From the quasi-objective standpoint, biogenesis has been treated from the viewpoints of irreversible thermodynamics (8) (11) (12), information theory (4) (16) or even by combining these with the theory of selection (3) (7). Of special importance in this respect is the paper of EIGEN that recently appeared elsewhere (5). How valuable these approaches, especially the last mentioned, ever may be, they are nonetheless able to illuminate the problem from one side only, which by necessity must be supplemented by another one in which is sufficiently enough demonstrated how strongly scientific propositions, mainly those of general character, are influenced by the mode of description (or, even, by the mode of cognition). Indeed, this point has been emphasized by two important papers in this volume (13) (14). This problem ultimately amounts to the question of what must happen in the human mind that enables an origin, i.e. the emergence of the entirely new, to be recognized, and, in addition, treated according to scientific criteria. One has to ponder on whether the recognition of an origin is the result of an active mental construction during the performance of cognition.

As has been shown, "the acquisition of skill", when taken as a paradigm for genesis as such, has "little, if any verbal content" (1). This has to be understood in the sense that the context (i.e. the meaningful "global" frame) must be given prior to the relations framed in it. In order to achieve a knowledge of that very context it is necessary, that after having started by considering the relations and the elements embedded in them (i.e. the "local", microscopic degrees of freedom) one has deliberately to change the mode of description in order to arrive at the larger "global" scope or realm. This process would be simply tantamount to saying that knowledge is an active seizing upon reality and brought forward only by appropriately applying the constructive principles inherent in mind. Such a principle is invoked by showing how complexification in one aspect leads to simplification in another, or vice versa (13) (14), whereby in active handling with the given reality new characters can be established, i.e. recognized. The statement has been made - which is of principal significance in this respect - that by means of a quantitative increase, i.e. increase in redundancy, the qualitatively new appears as the consequence of the action of new correlations or constraints (13) (14). This is, indeed, an expression for the reflexive or descriptive capability of the human mind to deal with different levels of reality. This can only be done by stating an atemporal form or pattern into which the temporal process is inserted in order to understand it. There are reasons that speak in favor of arguing that the complexification (or the dissipation) occurs at the lower level of description, whereas the correlation, i.e. the simplification, is brought about at the higher level. The statement mentioned above shows up as an assumption underlying the scientific theory (5) when there is said that the build-up of a new steady state is caused by the breakdown of a former steady state. And the same conception can be refound in the purely scientific representation of biogenesis when it is assumed that "the nucleation of functional correlation (we may call it the origin of life)" occurs out "of potential coupling factors" (5).

Thus, it can be said that no scientific theory can describe reality in a pretended objective way unless the descriptive (including the metaphysical) principles are taken into account which underlie the particular theory. This holds especially true for any fundamental problem to which, of course, that of the origin of life belongs. Therefore it must be gladly underlined that also in one abstract-paper of the symposium the role of the observer is at least touched upon (4). We may quote here the theory of cognition propounded by MATURANA (10). According to this theory the living organism exhibits a circularity, for which "hypercycles" (5) could be taken as paradigms on the molecular level. But this circularity exists only by dint of stating it as invariant and self-identical through the observer. It is he who, together with the observed system and its pertinent environment, constitutes a closed domain that cannot be escaped. By stating or asserting the identity of the circularity the observer places himself in the position to record a change in its appearance. Any description of the origin or evolution of life as a result of an intricate mechanism (in which mutation, selection, "valued" information, etc. meet together), despite its scientific garb, remains on the surface of the problem. Since it is to be acknowledged that for being able to evolve (or, what is the same, being recorded as evolving) the system (circularity etc.) must be put into (or set equal to) a framework only in relation to which evolution as, e.g. increase of complexity, can be assessed, this framework is of prime importance. By performing descriptions (i.e. performing reflections) the framework must be constantly altered and expanded but in the ultimate resort it is the observer (together with his cognitive domain) (10) who enters every description (as framework) and in comparison to whom every described entity represents an analogue to himself.

By application of the notion of the "non-holonomic" constraint (13) the influence of the describer (i.e. the observer) on the described (i.e. the observed) can be exemplified via a state space description. Although, e.g. the nucleic acids obey quantum mechanics (description), the enzymes responsible for DNA replication, RNA synthesis, etc. cannot be deduced from the fundamental laws of quantum mechanics (description) but only from classical physics (description). If the object of the description plus the measuring device (i.e. observer) would be considered a single quantum mechanical system then the measuring process would not be recognizable (and the measurement could not be detected or carried out). The classical nature of the measurement as opposed to the quantum mechanical events is analogous to the simplification vs. complexification in the description.

We should, in addition, not overlook that the process of evolution, if it should aim at the achievement of goals, e.g. an optimal adaptation to the environment or "survival of the fittest", is solely dependent on the pre-assigned system's laws and upon a kind of permanent but necessary interference with this very process (2). What is considered in nature as objectively occurring selection must in a formally contrived self-organizing system be done by the observer who, formally, can be replaced by another system (2) (9), of which then the environment can be considered a part. If, as is stated in the recently proposed theory by EIGEN (5) evolution becomes an inevitable event, then probably, in order to satisfy this metaphysical principle, a simpler description of the process would suffice and the expenditure of a complicated theory would not be required. But by no means is any assumption proved by a theory that follows from it. This fact, however, can be concealed if already the assumption is given a scientific garb.

Any fine theory always tends to seduce one to forget that it is only really inevitably dependent on the observer and his mode of cognition (including his metaphysical assumptions). Every theory is ultimately nothing but a mirror of what has been previously assumed. In building the theory the prerequisite has to be made compatible with the descriptive means so that by starting from a certain assumption, in order to get insight into (i.e. to mentally perform) a process, an opposite assumption must follow, in cancelling the primary one, and so on. If the metaphysical assumption is made that at the beginning there was a molecular chaos, then, in agreement with this assumption, a random start appears possible (5); but in order to continue, one must now by necessity make allowance for a kind of correlation, etc. However, the metaphysical postulate of the primary molecular chaos, although so-called "facts" seem to speak in favor of it, offers no advantage over an opposite principle. It is equally justifiable to postulate, in conformity with the order and ordering capability of the human mind, an opposite principle dominating the beginning, as has obviously been done also in this volume (7). Then the occurrence of evolution must be interpreted as the continuous process of dissolution (dissipation) and reforming of order, which again can only be done owing to the descriptive means. Even both initial statements could be combined in cogitating about the origin as arising out of an interplay between order and disorder (6).

Whenever a theory is formed it assumes, according to its presuppositions, a certain place within the realm of the reflective and descriptive means. The primary supposition of an objectively given matter is soon left when a formal approach can be made to the description of processes and the delineation of special and verifiable mechanisms. The mechanistic statement in turn will be cancelled and embraced by a systems theoretical view so that what at first sight (and at a lower level of description) appeared as contingent could now reveal itself as obeying everlasting (formal) laws, thus (by passing through contingency) reaching the realm of necessity. The supreme state of a theory, however, is achieved when it can be shown that what seems to be recognized as objectively existing is actually created by the human mind. Indeed, by taking into account the dependence of every view on the presupposed assumptions (made via the guiding lines of the mode of cognition), one escapes self-delusion that something objectively existing could ever be recognized. In having accepted this as theory one will become aware that in the last resort, one is seeing, although in a peculiarily distorted way (15), only oneself.

References

1. ANDREW, A.M.: This Volume
2. ASHBY, W.R.: In: H. v. Foerster, G.W. Zopf (Eds): Principles of Self-Organization, New York, Pergamon Pr., 1962, p.255.
3. BREMERMANN, H.J.: This Volume.
4. DECKER, P.: This Volume
5. EIGEN, M.: Naturwiss. $\underline{58}$, 465 (1971).
6. FOERSTER, H.v.: In: Self-Organizing Systems, p. 31.
7. FONG, P.: This Volume.
8. GROSS, B. & KIM, Y.G.: This Volume.
9. IVAKHNENKO, A.G., IEEE Transact. AC-8 (3) 247 (1963).
10. MATURANA, H.R.: Biology of Cognition, BCL-Rep.No. 9.0, Dep. Electr. Engin. Univ.Illinois, Urbana, Ill., 1970.
11. MEL, H.C. & EWALD, D.A.: This Volume.
12. NICOLIS, G.: This Volume.
13. PATTEE, H.H., in: L.L. Whyte, A.G. Wilson & D. Wilson (Eds.): Hierarchical Structures, New York, Elsevier, 1969, p. 161; J. Theor. Biol. $\underline{17}$, 410 (1967); This Volume.
14. ROSEN, R.: This Volume.
15. SPENCER-BROWN, G.: Laws of Form, London, George Allen and Unwin, 1969.
16. YOCKEY, H.P.: This Volume.

Subject Index

Actuality 1, 144
adaptation 1, 3, 8, 35, 150, 156, 176
adjustment, network 153
affinity squared function 107, 108
aging 16, 96, 166
algorithm, brain 150
 --, global optimization 29, 32
 --, information transmission 22
 --, preselected 155
allometric law 129, 135
analog simulation, enzyme systems 67, 69
antinomy, linguistic 169 ff.
"archetype" 33
artificial intelligence 148, 157, 175
association unit 150
atemporality 181
attention mechanism 153
attractor, ecological 34, 36
 --, system's 29, 32 ff.
automation, finite, 150, 155
automorphism 5

Backman's time 135
basin, space 29
beat frequency 66
behavior, coherent 44
 --, oscillatory 66
 --, very complex 175
"biochemical soup" 94, 114
bioids 25
body size 127 ff.
brain activity 149, 160
 --, development 98
 -- model, mathematical 173 ff.

Catastrophe, evolutionary 106, 174
 --, error 16
 --points 30, 34
 --, system's 30, 33
categories, mathematics of 139
"central dogma" 102
chalone 55
channel capacity, theorem 16

character, functional 115
chronometry, biological 131
circularity 182
cistron 10
classification, descriptive 178
clock, biological 131
clustering 3, 4
coacervate model 88
code, degeneracy 122
 --, doublet 15, 20
 --, one-to-one 9, 12, 18
 --, origin of 44, 46
 -- redundance 15
 --, triplet 21
codon frequency 12, 13, 21
cognition, theory of 181, 182
collagen 11
compartmentalization 63
competition, evolutionary 97
completeness postulate 42, 138
complexity 1 ff.
complication, principle of 99
computer simulation, evolutionary systems 140 ff.
 -- --, oscillatory systems 66, 76, 81, 82, 105
 -- --, neural networks 152, 153, 176
concentration waves 110
concept, forming of 150, 171, 179
conformation control 53
connectivity, neuronal 159
constraint, adaptive 11
 --, classifying 119
 --, interactive 36, 48
 --, optimization of 39, 45, 48
 --, system's 1 ff., 61, 108, 119
constructivity 144, 171, 174, 181
"constructive space" 138
context 7, 181
 --, dependence 155, 172
contraction, frequency 78
control, cellular 11
 --, code 8
 -- device 42
 --, mechanism of 98

control, metalogical 171, 172
--, optimal 32
--, origin of 41
--, proliferation 55
-- theory 61, 172
--, system's 119
-- units 71
conversion, principle of 99
--, discrete-continuous 160
cooperation, evolutionary 97
coordination, relative 135
cooscillations 108
Cope's law 127, 133, 135
correlations, functional 133, 135
--, generation of 40, 116, 119
correspondence principle 163, 179
coupling, elements 2
--, oscillators 74
--, system-environment 115
-- variable 112
cost function 113
-- value 59
creativity 5
critical points 32
cross-coupling 53

Death 96, 97
decision 4
-- making component 59, 121
decomposition, inputs 154
--, optimal 61, 62, 117, 119
degeneracy 10, 21, 122
degree of freedom 34, 42 ff., 74, 86, 181
--, redundant 116 ff.
description, alternative 41 ff., 119
--, complementary 48
-- levels 114
-- means 182
--, microscopic 42, 47
--, mode of 2, 4
--, role of the observer 181, 182
detail, loss of 41 ff., 178
determinism, local 35, 37
differentation, principle of 97
diffusion 64, 67, 68, 116
dimension, biological functions 128, 130, 134
dimensionality 116, 169
"directive correlation" 157
discrimination, maximal 140, 145
dissipative process 43 ff., 87, 109, 111
discrete processes 79
disturbance, v. perturbation
diversity measure 11
DNA 9, 10, 102

DNA, circular 125
-- evolution 122
--, information content 12
--, information storage 98, 102
--, memory storage 161
--, repair 96
--, replication 55, 94
--, selective advantages 125
dynamics, probabilistic 31
dynamic programming 111

Efficiency, organizational 98, 104
emergence, living system 4, 62
-- theory 91, 114
energy exchange, oscillatory 81
-- system-environment 131
-- flow 103
entrainment s. synchronization
--, 80 ff.
--, range of 82
--, relative 82, 83
entropy content 12
-- vs. information 11, 85, 91, 104
-- production, maximum 103, 104
--, minimum 108
--, misunderstandings 11, 86
--, probability distribution 11
--, system's 88
environment, change 36 ff.
--, factors 31
--, fluctuating 123
--, structure of 45
enzyme kinetics 51
-- reactions, autocatalytic 63 ff., 69
"error catastrophe" 16
errors, genome 12
--, recording 13
evaluator 58
event-memory system 139
evolution, chemical 88
--, mechanism of 97
excitability 46, 85, 90

Far-from-equilibrium reactions 107
feedback control 80
--, positive 69
-- system 107
fitness, environmental 25, 121
--, population 39, 40
flexibility, evolutionary 35, 36
fluctuation, information source 95, 105
fluid dynamics 110
form constraints 35
-- exaggeration 135

form, topological 33
-- transformation 113, 118
formalism, dynamical 34
formation, spontaneous 86
function, specificity of 45
functional change, principle of 114, 115, 121, 122

General systems theory 144
generation, genetic specificity 17
--, novelties 114, 115
--, state variables 115 ff., 122
genome, information content 18
glycolysis 53
glycoproteins, membrane-bound 165, 166
goal achievement 150, 152, 157
-- directedness, evolutionary 108
-- --, neural networks 152, 156
-- dynamics 32
-- evaluation 59, 157
-- generation 62
-- imposition 61
Goodwin system 107, 111
gradient field 30
gradualness 44, 45

Heart rate 129
heat production 127
hemoglobin, entropy 11
--, evolution 17
--, growth rate 166
hetero-pair 172
heuristic 157
hierarchy, biosystem 27
--, control 119
--, description 115
-- levels 44, 48
--, origin of 71
--, temporal 139, 144
history, developmental 118
--, life processes 94
holistic 148
holography 160
hypersurfaces, probabilistic 33

Identification problem 2
imperfection, principle of 98, 102
information accumulation 97, 102
--, bound 86, 91
-- change 97, 99, 102
-- content 18, 21, 93, 95
-- creation 94
--, definition 102
--, dissipation 16, 94 ff., 103

information, environmental 88
--, generation 21
--, kinds of 25, 147
--, meaning 11
--, positional 55
-- processing, nervous system 148, 155
--, quality 91, 122
-- retrieval 98
-- safe-keeping 96
-- theory 3, 9, 10, 21, 85
-- transformation 94
-- transmission 13, 85, 102, 104
input-output relationship 149, 160
instability 35
interaction, cellular 116
--, descriptions 115, 171, 177
--, oscillatory 73 ff.
--, simplicity-complexity 2, 4
--, structure-function 114 ff., 121
--, systems 46, 117
--, system-environment 91, 115
--, transfer function 59, 60
interpretation, semantic 179
intuitionism 144
invariance 8

Juxtaposition principle 97, 100

Kinetics, oscillatory 63, 74 ff.
--, quantities 109

Language, antinomy 169, 170
--, programming 170
-- structure 148
--, theory of 169
--, use of 148
--, value of 44 ff., 104
"learnability" 176, 177
learning filter 153
--, genetic 155, 156
-- system, neuronal 148, 159, 166, 175
levels, description, 114, 171
-- membership 140
--, organizational 131
--, quality 116
life, characteristics 95
-- expectancy 135
light action, oscillators 77
limit cycle 29, 67, 68, 77, 107
--, data processing 33
limitation, system capabilities 150
linearity hypothesis 174, 177
logic, heterogeneous 169, 172
--, many valued 171

lumping processes 46, 49, 117, 175

Manifold, differentiable 138
mapping, intrasystem 2, 3, 4
--, environment-system 25
maximum principle 111
McCullogh-Pitts model 157
-- -- theorem 7
meaning, biochemical 18
--, informational 22, 104, 105
--, symbols 104
measurement 3, 43
membrane potential 85
-- structure, change 165
memory, finite automaton 150
-- hierarchy 140, 144, 145
--, long-term 159, 166
--, mathematical analogies 160, 161
--, models 163, 176
--, short-term 165
-- structure 144
--, switching theory 161
Mendelian inheritance 141
message, genetic 16
metabolic reduction, law of 127, 135
metalanguage 48
metaorganization 48
Michaelis-Menten law 77
minimization principles 107, 108, 111, 113
--, trajectories 113
mode of cognition 181
model, biochemical 76
--, brain, algebraic 172 ff.
-- --, topological 172 ff.
--, compartment 64 ff.
--, enzyme 63, 66, 67
--, learning 163
--, memory 163
--, one-language 178
--, organism 113
--, notion of 47, 128
--, relational 71
--, test of 177
morphogenesis 30, 33
morphology, characteristics 113, 120
motion, equations of 27
multilayer adjustment 153
multiplication, exponential 94
mutation 9, 15, 31, 95
--, multiplet 31

Negentropie 122
network, metabolic 71
--, principle of 96

neural net 151, 152
-- tissue 85
neuron model 149
--, physical 171, 179
neutron capture 87
-- scattering 87
niche, ecological 33, 36
noise accumulation 22
--, genetic 16
--, signal 152, 155, 156
non-reductionism 43
nucleation matrix 86

Observation, constituents 119
--, criteria 119
--, macroscopic 73
observer 3, 85
--, change 5
--, constructive center 139
--, information 26
--, mutual 120
ontogenesis, neural 165
operon 55
optimality, evolutionary 39, 40
--, local 33, 37
optimization constraints 39
--, evolutional 32, 91
--, global 32, 60, 121
--, random directions 32, 33
order disorder, interplay 183
-- emergence 110
--, hierarchical 1, 2, 4
organization, cellular 53, 86
--, definition 103
--, levels 131
--, precellular 88
origin, goals 156
--, hierarchies 115, 119
--, oscillatory systems 80
--, problem 6, 18, 27, 43, 94
--, recognition of 181, 183
oscillation frequencies 70
--, light frequency 77
--, synchronized 73 ff., 79
--, nonlinear 76, 77
--, neural 79
--, relaxation 131, 134, 135
oscillators, interacting 73
--, monofrequency 75
--, undamped 66, 67
oxidation, photosensitized 71

Pace maker activity 131, 135
parameterization, memory 161

pattern recognition 155
pendulum oscillation 81, 131
perceptron 150, 156
perfection, principle of 98, 102
performance function, evolutionary 39
periodicities, biological 130
perturbation 2, 30, 34, 35, 37, 41, 67, 68, 71, 77, 80, 81, 161
phase space, zonation 76
polyphosphate complexes 88, 89, 125
potential function 31
--, thermodynamic 108
potentiality 1, 6, 144
power law, biological time 133
precursor 64 ff.
"primordial soup" 94
Pringle's criterion 3, 4
principles, evolutionary 96 ff.
--, selforganization 148
--, uncertainty 62
process description 120
product space 1
program, system's 1
proliferation 116, 121
protein synthesis 55, 102
"prototype" 128
P-structure 1, 2
punctuation 12, 21
purposiveness 3, 99, 102, 147

Random fluctuation, enzyme model 68
read-out mechanism 159, 162, 163
realizability, physical 69
realization 4, 27
record, error 13
--, genetic 47
--, systems 4
reducibility, descriptive 41, 42
redundance, descriptive 2
--, error correction 12
-- increase 116
--, informational 10
-- vs. noise 152, 155
--, principle of 96, 100
--, reduction of 150, 152, 156
--, relational 27
--, spatial 117
relation, 1 ff., 27, 48
-- coordinative 169
"relational biology" 27, 118
relaxation 86
--, oscillation 81
repair 8, 96
representation 4

representation, theorem 12
reproduction 99
revolution, principles 100
rigidity, mathematical 173 ff.

Selection, interaction transfer function 61
--, natural 9, 39, 40
self 4, 5, 48
-- adaptation 111
-- analog 172
-- coupling 53
-- control 111, 112
-- organization 3, 147 ff., 152, 155, 157
-- realization 4
-- reference 170
-- repair 8
-- representation 4
-- reproduction 103
sensitivity analysis, automatic 151, 156
sex 96, 99
significance feedback 147, 150 ff.
similarity, biological 129, 131, 134, 140
--, formal theories 113
--, mechanical 128, 134
simplification 4, 43, 74, 178, 181, 182
simulation, analog computer 66, 69
space, cellular 53
-- consideration 138
-- , constructive 144
-- , extension 135, 138
-- , oscillation 77
-- , traveling 143
specialization 99
specificity, biochemical 18
-- , genetic 17
spontaneity, evolutionary 103
S-structure 1
stability asymptotic 103, 117
--, conditional 34, 35
-- criterion 107
--, nonlinear equations 161
--, structural 6, 30, 34
--, system's 27, 31, 41
-- theory 109
state description 120
-- function 103
-- space 34, 37, 182
--, synchronous 73, 79
-- transition 107
stimulus-response 3, 8
structure, brain 149
--, dissipative 109
--, hierarchical 5, 139
--, mathematical 138, 144

structure, metastable 88
structuralization 87, 90
substrate inhibition 69
substitution, principle of 97
superposition, state 120
survival 88
symbols 48
symmetry breaking 177
synchronization 73, 76, 131
--, mutual 77, 80
system, adaptive 176
  -- activity 1
  -- analysis 2
  -- classification 3
  -- connectivity 7
  --, conservative 35
  -- decomposition 46, 61
  -- description 31, 46
  -- --, levels 114
  --, dynamical 29, 35, 43, 46, 74, 80
  --, experimental 60
  -- generation 4
  --, hierarchical 115
  -- inconsistencies 97
  --, linear 76
  -- organization 2, 8, 62
  --, nonlinear 60, 76, 78
  --, real 118
  --, structure 63, 66
  --, synergic 57, 60, 61, 121
  -- synthesis 2
  --, teleogenic 57
  -- theory, mathematical 175
  -- variables 1, 3

Teachability 176
teleogenesis 57, 62, 156
temperature action, enzymes 63 ff.
  --, measurement 11
template, first molecule 20
temporality, classical 138
theories, structural 115
"thermodynamic potential" 107
thermodynamics 3, 43
  --, irreversible 109, 110
  --, second law 96, 108
thresholdelement 149, 175
  --, neuronal, change 159
time, epochal 145

time, local-global 135, 144
  -- measurement 128
  --, objective 144
  --, operational 5, 105, 130 ff., 144
  --, physiological 131, 133
  , real 113, 118
  -- reversal 139
  -- scale 5, 118, 135, 144
  -- structure 139
  -- travel 143, 144
timelessness 135, 137 ff.
topology 29
Turing system 103, 116, 117, 122
  -- machine 169, 179
training algorithm 151
trajectories, averaged 112
  --, breaking 30
  -- parameterization 118
  --, periodic 77
  --, process 31, 32
  --, system 34 ff.
transcomputability 32, 37
transformation theory 91, 113 ff.
transient time 53
transition, descriptive 98
  --, oscillatory modes 66, 74 ff., 81
  --, societies 99
  --, species 113
  --, state 2, 3, 33, 107

Utility function, global 59, 60, 121
uv irradiation, low level 71

Volterra-Lotka system 27, 67
value definition 36, 85, 86
  --, elective 114
  --, minimum, constancy 108
  --, observable 120
  --, worth assigning 58, 121
Van der Pol oscillator 81
variable, sensitive 43
versatility, principle of 99

Wilkinson's law 135
worth functional 60
  -- value 58

Zeitgeber 81, 83

Springer-Verlag
Berlin · Heidelberg · New York
München London Paris Sydney Tokyo Wien

Quantitative Biology of Metabolism

Models of Metabolism
Metabolic Parameters
Damage to Metabolism
Metabolic Control

Organizers:
A. Locker, F. Krüger
Editor: A. Locker

3rd International Symposium
Biologische Anstalt Helgoland
September 26-29, 1967

With 66 figures
XVI, 296 pages. 1968
Soft cover DM 70,–; US $ 25.90

Prices are subject to change
without notice.

■ Prospectus on request!

This Symposium carries on the attempts of those held in 1963 and 1965 to introduce abstraction into the unwieldy mass of experimental biological data so that it can be handled in theoretical i.e. mathematical terms, for only on in this way can biology advance from the "perscientific" to the truly scientific stage.

A variety of systems-theoretical considerations and of models are applied to problems of cellular metabolism, the models being discrete or continuous, deterministic or stochastic, as appropriate. Models based on different mathematical statements are compared and the value of computer simulation is demonstrated.

Unlike metabolic processes as such, the influence of metabolic parameters (e. g. temperature, season, adaptation and to some extent ageing) does not lend itself to treatment by mathematical models but must for the time being continue to be described quantitatively. A striking exception is growth, for which mathematical models have long been proposed. An effective application of mathematical treatment is provided by experiments on radiation damage, goth reversible and irreversible, the means by wich the organism seeks to resist injury or promote recovery exemplifying "optimization" in metabolic control, with modifications occurring at different levels of biological organization.

Journal of Molecular Evolution

Research in widely different fields is converging into what may properly be called the new science of molecular evolution. The recently established JOURNAL OF MOLECULAR EVOLUTION makes possible the ready availability of the increasing amount of relevant research in this field.

One Volume (4 issues) is published annually
DM 136,–; US $ 50.40

**Springer-Verlag
Berlin
Heidelberg
New York**
München London Paris
Sydney Tokyo Wien

This international journal covers the following areas:
1. biogenetic evolution (prebiotic molecules and their interaction)
2. evolution of informational macromolecules (primary through quaternary structure)
3. evolution of genetic control mechanisms
4. evolution of enzyme systems and their products
5. evolution of macromolecular systems (chromosomes, mitochondria, membranes, etc.)
6. evolutionary aspects of molecular population genetics

Editor-in-Chief:
Dr. E. Zuckerkandl, Département de Biochimie Macromoléculaire,
B. P. 5051, 34033 Montpellier Cedex, France

Editors:
G. Braunitzer, Munich
R. E. Dickerson, Pasadena, Calif.
J. L. King, Santa Barbara, Calif.
C. Ponnamperuma, College Park, Md.

Editorial Board:
E. A. Barnard, Buffalo, N.Y.
E. S. Barghoorn, Cambridge, Mass.
R. J. Britten, Washington, D. C.
R. Buvet, Paris
M. Calvin, Berkeley, Calif.
Ch. R. Cantor, New York, N.Y.
L. L. Cavalli-Sforza, Palo Alto, Calif.
J. F. Crow, Madison, Wis.
M. O. Dayhoff, Washington, D. C.
G. H. Dixon, Vancouver
R. F. Doolittle, La Jolla, Calif.
R. V. Eck, Athens, Ga.
W. M. Fitch, Madison, Wis.
M. Florkin, Liège
N. H. Horowitz, Pasadena, Calif.
R. T. Jones, Portland, Oreg.
Th. H. Jukes, Berkeley, Calif.
N. O. Kaplan, La Jolla, Calif.
M. Kimura, Mishima
H. Lehmann, Cambridge
B. McCarthy, Seattle, Wash.
M. Mandel, Houston, Tex.
E. Margoliash, North Chicago, Ill.
Zh. A. Medvedev, Obninsk
J. V. Neel, Ann Arbor, Mich.
H. Neurath, Seattle, Wash.
A. I. Oparin, Moscow
L. Orgel, London
J. Oró, Houston, Tex.
L. Pauling, Big Sur, Calif.
J. F. Pechère, Montpellier
G. R. Philipps, Bonn
B. Pullman, Paris
Ch. Sadron, Orléans
E. Schoffeniels, Liège
E. L. Smith, Los Angeles, Calif.
J. Maynard Smith, Brighton
S. Spiegelman, New York, N.Y.
G. L. Stebbins, Davis, Calif.
J. H. Subak-Sharpe, Glasgow
N. Sueoka, Princeton, N. J.
T. Swain, Kew, Surrey
B. L. Turner, Austin, Tex.
L. van Valen, Chicago, Ill.
H. J. Vogel, New York, N. Y.
P. M. B. Walker, Edinburgh
A. C. Wilson, Berkeley, Calif.
C. R. Woese, Urbana, Ill.

Authors are invited to submit manuscripts in English to the Editor-in-Chief.

Prices are subject to change without notice.